Sediments Contamination and Sustainable Remediation

Catherine Mulligan
Masaharu Fukue
Yoshio Sato

Publishing

CRC Press
Taylor & Francis Group
Boca Raton   London   New York

CRC Press is an imprint of the
Taylor & Francis Group, an **informa** business

Co-published by IWA Publishing, Alliance House, 12 Caxton Street, London SW1H 0QS, UK
Tel. +44 (0) 20 7654 5500, Fax +44 (0) 20 7654 5555
publications@iwap.co.uk
www.iwapublishing.com
1-84339-300-X
978-1-84339-300-9

CRC Press
Taylor & Francis Group
6000 Broken Sound Parkway NW, Suite 300
Boca Raton, FL 33487-2742

© 2010 by Taylor and Francis Group, LLC
CRC Press is an imprint of Taylor & Francis Group, an Informa business

No claim to original U.S. Government works

Printed in the United States of America on acid-free paper
10 9 8 7 6 5 4 3 2 1

International Standard Book Number: 978-1-4200-6153-6 (Hardback)

Library of Congress Cataloging-in-Publication Data

Mulligan, Catherine N.
    Sediments contamination and sustainable remediation / authors, Catherine N. Mulligan,
Masaharu Fukue, and Yoshio Sato.
        p. cm.
    "A CRC title."
    Includes bibliographical references and index.
    ISBN 978-1-4200-6153-6 (Taylor & Francis : alk. paper) -- ISBN 978-1-84339-300-9 (IWA :
alk. paper)
    1. Contaminated sediments--Management. 2. Soil remediation. 3. Sedimentation and
deposition. I. Fukue, Masaharu. II. Sato, Yoshio, 1947- III. Title.

TD878.M85 2010
628.1'68--dc22   2009038395

Visit the Taylor & Francis Web site at
http://www.taylorandfrancis.com

and the CRC Press Web site at
http://www.crcpress.com

Contents

Preface

The surface water environment is an important part of the geoenvironment. It is the recipient of (a) liquid discharges from surface runoffs, rivers, and groundwater and (b) waste discharges from land-based industrial, municipal, and other anthropogenic sources. It is also a vital element that provides the base for life support systems and is a significant resource. The combination of these two large factors, with their direct link to human population, makes it an integral part of the considerations on the sustainability of the geoenvironment and its natural resources. A healthy ecosystem ensures that aquatic plants and animals are healthy and that these do not pose risks to human health when they form part of the food chain. In this book, we will discuss (a) the threats to the health of the sediments resulting from discharge of pollutants, excessive nutrients, and other hazardous substances from anthropogenic activities, (b) the impacts observed as a result of these discharges including the presence of hazardous materials and the phenomenon of eutrophication, (c) the remediation techniques developed to restore the health of the sediments, and (d) how to evaluate the remediation technologies using indicators. Therefore, the problem of sediment contamination is developed, in addition to how the sediments can be remediated and how the treatments can be evaluated.

Contaminated sediments are a risk to fish, humans, and animals that eat the fish. Although part of the geoenvironment, sediments have received much less attention from researchers, policy makers, and other professionals than other components. Sediment, however, is an essential and valuable resource in river basins and other aqueous environments. A large biodiversity exists in the sediments. It is thus a source of life and resources for humans as construction materials, sand for beaches, and farmland and wetland nutrients.

There is a need to develop a better understanding of the sediment–water environment and better management practices due to their potential impact on human health and the environment. In particular, they need to be considered during efforts to meet sustainability requirements. Sediments can be exposed to multiple sources of contaminants and are located at the bottom of water columns. This makes risk assessment and management more difficult than in soils. The benthic community cannot be isolated from the contaminated sediments. This community is at the base of the aquatic food chain, but can be highly tolerant to the contaminants. Sediment quality criteria thus are much lower than for soils because the sediments can have a significant influence on the aquatic food chain.

Sediments have been removed for centuries by dredging for maintaining navigation. This type of sediment management will not be elaborated on substantially because sediment management for the purpose of environmental cleanup or management will be the main focus of this book. The binding of the contaminants to the sediments, their bioavailability, mobility, and degradability are all important aspects that will be taken into account. More than 10% of the sediments have been estimated to be contaminated in the United States.

In Chapter 1, we will focus on the introduction of the importance of sediments, the sources of contaminants, management practices, and sustainability. Sediments are found in lakes, rivers, streams, harbors, and estuaries after traveling downstream from their origin. Sources of effluents containing the solids include urban, agricultural, and industrial lands. Strategies for remediation of contaminated sediments are introduced.

In Chapter 2, sediment components are discussed. They are inorganic and organic solid materials and are often classified by size, as gravel, sand, silt, and clay. The term "sediments" is used for soils deposited in water. They are often called marine soils, if it is settled in the marine environment. Thus, sediments are in contact with inorganic, organic, and other human-discharged materials, through the influence of the pore water. Therefore, the properties of the pore water are an important factor regarding the quality of the sediments. Sediment uses are also described.

The interactions of the pore water with the contaminants and the solids are complex and will be discussed in Chapter 3. It is important to understand the physical and physicochemical interactions of the contaminants with the sediment solids to understand the capacity of retention of the sediments and potential parameters for contaminant release. Sediment composition, properties, and characteristics will influence the interactions at the sediment–pore water interface. The reactions between pollutants and sediment will determine its transport through the sediments, and also its fate.

Sediment quality is related to the quality of surface water. It is due to the serial mechanisms of the dissolution of organic matter and the exclusion of contaminants due to the consolidation of sediments or the leaching of contaminants. Therefore, in order to make an appropriate assessment of sediments, the physical, chemical, and biological mechanisms have to be understood well. Since the mechanisms are natural and complex, there is the possibility that nonpredictable results can be obtained. Therefore, it is necessary for engineers to modify or take measures suited to the occasion.

In Chapter 4, information including sampling and physical, chemical, and biological test procedures to determine the state and extent of contamination will be examined. Sampling can also be used to predict future trends or to evaluate the progress of the remediation work. The scale for sampling and monitoring will be site dependent. Since most of the physical and chemical properties of sediments have to be determined by the laboratory tests, sampling is almost always needed. Therefore, monitoring of sediment properties can be achieved by tests on samples obtained from the sites. Thus, much effort and planning is required for the monitoring of sediments.

In Chapter 5, the mechanisms involved and case studies of natural recovery of various pollutants at contaminated sediment sites will be examined. There are differences in the type of processes that play a role in the natural attenuation of groundwater and the natural recovery of sediments. Usually transformation processes of the contaminants are more dominant in the natural attenuation of surface soils, whereas isolation and mixing are more prevalent in sediments. Natural recovery includes both attenuation aspects (reduction of contaminants with no transport to other media) and recovery (which allows the benthic and pelagic communities to be reestablished and resume their beneficial uses). Monitoring is required to ensure that the remediation objectives are achieved and that it is proceeding as planned. Thus the term monitored natural recovery (MNR) is used. Thus upon successful completion, MNR would meet the needs of sustainability. Acceptance is increasing as there is substantial cost reduction achieved due to the nonremoval of large volumes of sediments.

There are still many gaps in knowledge, and a careful evaluation of the management options must be made. Techniques for the remediation of sediments may be required when the sediment leads to the accumulation of contaminants in aquatic life or when the release of hazardous materials from sediments becomes a serious problem. Therefore, a remediation technique, such as capping, dredging, or physical, biological, and/or chemical treatment, has to be considered. In Chapter 6, in situ remediation techniques and the management of contaminated sediments will be described. In situ remediation could be beneficial over dredging due to a reduction in costs and lack of solid disposal requirements. Therefore, these methods will be examined.

In Chapter 7, dredging and the management of dredged sediments will be discussed. Dredging is the excavation of materials (sediments) from the bottom of the water column for a number of different purposes and is often required for navigational purposes in coastal and inland waters and/or removal of contaminated sediment. The dredging process itself has the potential to impact the environment. Proper design of the dredging project can minimize the environmental impact. Long-term monitoring is rarely performed to determine the residual contamination and long-term effects of the dredging. The use of different methodologies includes physico-chemical to biological approaches to the management of different routes of disposal or uses of the dredged material.

Selection of the most appropriate remediation technology must coincide with the environmental characteristics of the site and the ongoing fate and transport processes and is elaborated on in Chapter 8. To be sustainable, the risk at the site must be reduced, and the risk should not be transferred to another site. The treatment must reduce the risk to human health and the environment. Cost-effectiveness and permanent solutions are significant factors in determining the treatment. Sites vary substantially, and there can be substantial uncertainty involved in the evaluation process. However, decisions must be made based on the information available. In this chapter, we will examine the means to select the most appropriate technique for site remediation, evaluate the progress of the remediation, and determine the long-term restoration of the site.

Finally, in Chapter 9, the two main approaches, in situ and ex situ treatment, are examined further. Environmental dredging requires evaluation of the risks of dredging, determination of disposal methods, and/or potential beneficial use. Depending on site conditions, in situ management may be preferable and may pose less risk to human health, fisheries, and the environment. Both short-term and long-term risks must be evaluated for the in situ and ex situ options. To work toward sustainability, waste must be minimized, natural resources must be conserved, landfill deposition should be minimized, and benthic habitats and wetlands must not be lost and must be protected. Innovative integrated decontamination technologies must be utilized. We will examine, also, where developments are needed. The fate and transport of contaminants must be understood more thoroughly to develop appropriate strategies. Sediment quality standards and guidelines and conventions are detailed in the appendices.

We wish to acknowledge the benefit of all the interactions and discussions we have had with all colleagues, research students, and professionals in the field. They are all a vital part of the education of the public, industry, and governmental bodies that are involved in the conservation and protection of the natural resources. A long-term vision is needed. Otherwise, natural resources will continue to be depleted, landfills will continue to be filled with contaminated sediments, and biodiversity in the aquatic geoenvironment will be diminished. Integrated innovative management practices need to be developed and applied.

Catherine N. Mulligan
Masaharu Fukue
Yoshio Sato

The Authors

Catherine N. Mulligan has B.Eng. and M.Eng. degrees in chemical engineering and a Ph.D. specializing in geoenvironmental engineering from McGill University, Montreal, Canada. She has gained more than 20 years of research experience in government, industrial, and academic environments. She was a research associate for the Biotechnology Research Institute of the National Research Council and then worked as a research engineer for SNC Research Corp., a subsidiary of SNC-Lavalin, Montreal, Canada. She then joined Concordia University, Montreal, Canada, in the Department of Building, Civil and Environmental Engineering. She has taught courses in site remediation, environmental engineering, fate and transport of contaminants, and geoenvironmental engineering, and she conducts research in remediation of contaminated soils, sediments, and water. She holds a Concordia Research Chair in Geoenvironmental Sustainability. She has completed a textbook (*Environmental Biotreatment*, Government Institutes, 2002) as a sole author on biological treatment technologies for air, water, waste, and soil, and two others, with Professor R.N. Yong (*Natural Attenuation of Contaminants in Soil*, CRC Press, 2004) and with Professors R.N. Yong and M. Fukue (*Geoenvironmental Sustainability*, CRC Press, 2006). She has authored more than 50 refereed papers in various journals and holds three patents. She is a member of the Order of Engineers of Quebec, Canadian Society of Chemical Engineering, American Institute of Chemical Engineering, Air and Waste Management Association, Association for the Environmental Health of Soils, American Chemistry Society, Canadian Society for Civil Engineering, and the Canadian Geotechnical Society.

Masaharu Fukue has B.Eng. and M.Eng. degrees in civil engineering from Tokai University, Japan, and a Ph.D. in geotechnical engineering from McGill University, Montreal, Canada. He joined a consultant firm for a short period and then moved to Tokai University. He has given courses in geoenvironmental engineering, hydrospheric environment, shipboard oceanographic laboratory, and submarine geotechnology. He is a member of the International Society for Soil Mechanics and Geotechnical Engineering, International Society for Terrain-Vehicle Systems, Japanese Society for Civil Engineers, the Japanese Geotechnical Society, and the Japan Society of Waste Management Experts. He served as a chief editor for *Japanese Standards for Soil Testing Methods* and for *Japanese Standards for Geotechnical and Geoenvironmental Investigation Methods*. He also served as a director of the Standard Division and a member of the Board of Directors of the Japanese Geotechnical Society. Since 1996 he has been a member of the editorial board of the American journal *Marine Georesources and Geotechnology*. He was recently a chair of the organizing committee of the 3rd International Symposium on Contaminated Sediments, sponsored by ASTM, ISCS2006-Shizuoka, Japan. He is also an advocate and a promoter of the Annual Symposium on Sea and Living Things and Rehabilitation of Coastal Environment, Japan. He has sponsored the Marine

Geoenvironmental Research Association in Japan. He invented a filtration system for seawater using a 2500-ton large barge and performed a field experiment using the system for seawater purification in a small bay. He has published more than 300 scientific papers on qualities of seawater, sediments, and soils, has completed a book with Professors R.N. Yong and C.N. Mulligan, *Geoenvironmental Sustainability* (CRC Press, 2006), and is co-editor of *Contaminated Sediments: Evaluation and Remediation Techniques*, STP 1482, ASTM International, 2006.

Yoshio Sato has B.Sci. and Dr.Sci. degrees in oceanography from Tokai University, Japan. He has had more than 30 years of research and teaching experience in the university. His specialty is chemical analyses of the formation mechanism of manganese nodules on the ocean floor. Recently, he has become interested in preservation of the environment of enclosed sea areas, the utilization of ground seawater for fishery, and deep ocean seawater. He is a member of the Chemical Society of Japan, the Oceanographic Society of Japan, the Geochemical Society of Japan, the Society of Sea Water Science, Japan, and the Japan Association of Deep Ocean Water Applications. He is a member of a committee for prevention of pollution in Shizuoka Prefecture, Committee of Environment in Shizuoka Prefecture, and Committee of Environment in Shizuoka City. He is a member of the board of directors of the Society of Sea Water Science, Japan, and has published more than 100 papers about seawater.

1 Introduction to Sediment Contamination and Management

1.1 INTRODUCTION

Approximately 0.9 billion m³ of sediment in the United States are contaminated, which are a risk to fish, humans, and animals that eat the fish, according to the United States Environmental Protection Agency (USEPA, 1998). The rate of survival, immunity to diseases, and growth of fish such as salmon may be affected by exposure to contaminated sediments early in life (Varanasi et al., 1993). Although part of the geoenvironment, sediments have received much less attention from researchers, policy makers, and other professionals than other components. Sediment, however, is an essential and valuable resource in river basins and other aqueous environments. A large biodiversity exists in the sediments. It is thus a source of life and resources for humans as construction materials, sand for beaches, and farmland and wetland nutrients.

However, due to the close contact of sediments with the water environment, they are both a source and a sink for contaminants. There is a need to develop a better understanding of the sediment–water environment and better management practices due to their potential impact on human health and the environment. In particular, they need to be considered during efforts to meet sustainability requirements. Some of the major impacts due to increasing population pressures include:

- Loss of biodiversity and living resources
- Increased production of wastes and pollutants
- Depletion of nonrenewable natural resources
- Decreased soil, water, and air quality
- Increased discharges of greenhouse gases

Although some of these issues have been examined previously in regard to the geoenvironment (Yong et al., 2006), in this book, we will focus on the stresses and how to mitigate the impacts of these factors in relation to sediments, because they are a highly important resource and basis for life. This environment will be defined as the aquatic geoenvironment (Figure 1.1). They form an integral part of a functioning ecosystem and partake in various types of physical, chemical, and biological activities. Some of these as detailed by Trevors (2003) include partaking in various cycles such as those of carbon, nitrogen, phosphorus, and sulfur, in addition to the

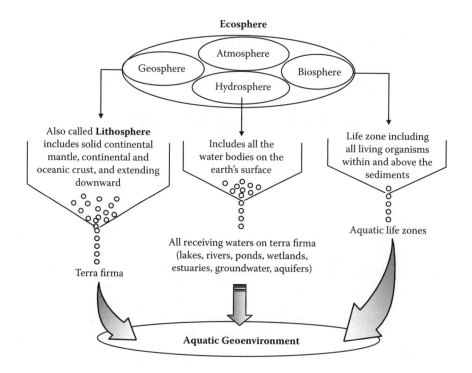

FIGURE 1.1 The various constituents of the ecosphere and their relationship to the aquatic geoenvironment.

hydrologic and natural processes for the control of the biodegradation of pollutants in the sediment and water.

Sediment is defined by SedNet as "suspended or deposited solids, acting as a main component of a matrix which has been or is susceptible to being transported by water" (Brils, 2003). Soil is defined as an aggregate material covering the earth surface which consists of solid particles and void spaces with liquid and gas. Soil particles are composed of inorganic and organic solid materials and are often classified by size, as gravel, sand, silt, and clay (which will be discussed in more detail in Chapter 2). The term "sediments" is used for soils deposited in water. They are often called marine soils, if they are settled in the marine environment. Sedimentary rock is, therefore, consolidated and cemented sediment.

Sediments can be exposed to multiple sources of contaminants and are located at the bottom of water columns. This makes risk assessment and management more difficult than for soils. The benthic community cannot be isolated from the contaminated sediments (USEPA, 2002). This community is at the base of the aquatic food chain, but can be highly tolerant to the contaminants (USEPA, 1998). Sediment quality criteria thus are much lower than for soils because the sediments can have a significant influence on the aquatic food chain. More than 10% of the volume of sediments (the upper 5 cm) at the bottom of the U.S. surface waters have been estimated to be contaminated.

1.2 SUSTAINABLE DEVELOPMENT AND THE AQUATIC GEOENVIRONMENT

Five major themes under the acronym sustainable development were identified in the Johannesburg World Summit on Sustainable Development (WSSD 2002). They included (1) water and sanitation, (2) energy, (3) health, (4) agriculture, and (5) biodiversity. It can easily be seen how many of these activities can influence sediment quality. The impact of development activities with four components can be substantial. The components include industrialization, urbanization, resource exploitation, and agriculture (food production) (Figure 1.2).

Sediments are found in lakes, rivers, streams, harbors, and estuaries after traveling downstream from their origin. Sources of effluents containing the solids include urban, agricultural, and industrial lands. Sediments have been removed for centuries by dredging for maintaining navigation. This type of sediment management will not be elaborated on substantially, because sediment management for the purpose of environmental cleanup or management will be the main focus of this book along with the assessment of the sediments. The binding of the contaminants to the sediments, their bioavailability, mobility, and degradability are all important aspects that will be examined.

1.3 SOURCES OF POLLUTANTS

Point and diffuse pollution sources enter the aquatic environment. Agricultural, urban, and industrial activities, spills, and accidents contribute to the pollution. Manufacturing and energy production, urban centers, municipalities, service industries, airborne and groundwater-transported contaminants all contribute contaminants to the sediments. In general, these effluents are either surface runoffs that

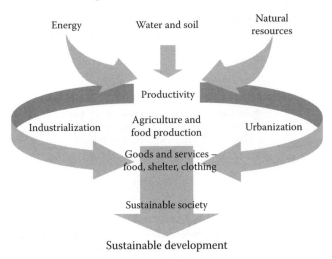

FIGURE 1.2 Basic elements and interactions contributing to a sustainable society and to sustainable development.

discharge into the rivers, lakes, and groundwater or are point sources (Figure 1.3) from municipal, industrial, or other sources.

Dredging is commonly used for maintenance of navigational routes. The material has been reused for building and construction materials. Extraction of oil and other resources is also frequent below the water surface, such as oil from Hibernia platforms of the coast of Newfoundland, Canada.

1. The use of the marine environment for fish and seafood extraction is one of the oldest industries. More recently, fish aquaculture is growing in popularity as fish stock become more and more depleted.
2. Water is extracted commonly for drinking water and for hydroelectric power generation.
3. Although waste disposal is most frequently on land, a lack of suitable land surfaces is now forcing waste disposal facilities in countries like Japan to be placed in marine landfills in special facilities.

Some of these contaminant sources and how they reach the marine geoenvironment can be seen in Figure 1.4. Proper management means that the impact must eliminate or minimize damage to the ecosystem and the entry of these pollutants into the environment. Spills, leaks, discharge, and runoff all threaten water quality and subsequently health, two of the main components of WEHAB.

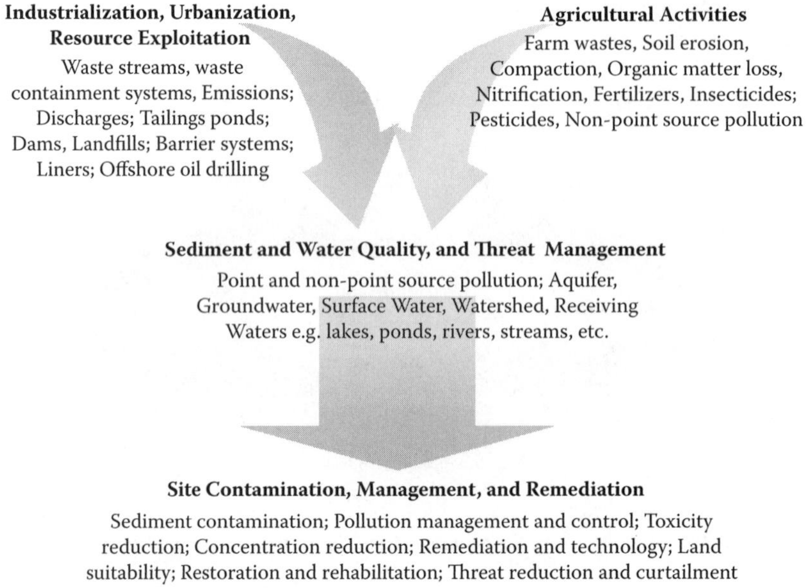

Industrialization, Urbanization, Resource Exploitation
Waste streams, waste containment systems, Emissions; Discharges; Tailings ponds; Dams, Landfills; Barrier systems; Liners; Offshore oil drilling

Agricultural Activities
Farm wastes, Soil erosion, Compaction, Organic matter loss, Nitrification, Fertilizers, Insecticides; Pesticides, Non-point source pollution

Sediment and Water Quality, and Threat Management
Point and non-point source pollution; Aquifer, Groundwater, Surface Water, Watershed, Receiving Waters e.g. lakes, ponds, rivers, streams, etc.

Site Contamination, Management, and Remediation
Sediment contamination; Pollution management and control; Toxicity reduction; Concentration reduction; Remediation and technology; Land suitability; Restoration and rehabilitation; Threat reduction and curtailment

FIGURE 1.3   Threats and waste streams impacting soil and water quality.

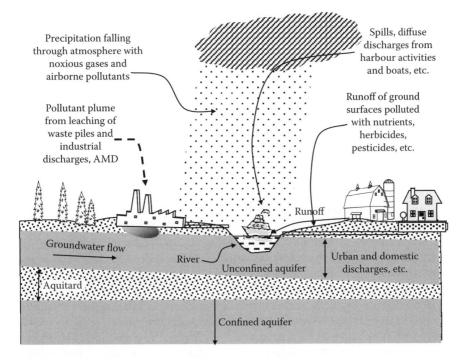

FIGURE 1.4 Some of the more prominent causes of pollution of recharge water for rivers, other receiving waters, and groundwater (aquifers). Contamination of the confined aquifer depends on whether communication is established with the unconfined aquifer.

Heavy metals are common inorganic pollutants in the geoenvironment. These include:

- From atomic numbers from 22 to 34: Ti, V, Cr, Mn, Fe, Co, Ni, Cu, Zn, Ga, Ge, As, and Se
- From 40 to 52: Zr, Nb, Mo, Tc, Ru, Rh, Pd, Ag, Cd, In, Sn, Sb, and Te
- From 72 to 83: Hf, Ta, W, Re, Os, Ir, Pt, Au, Hg, Tl, Pb, and Bi

Anthropogenic activities such as landfills, metal extraction, and metal plating generate heavy metal leachates containing copper, lead, zinc, and so on. A more detailed description of the various forms and sources of arsenic, cadmium, chromium, copper, lead, nickel, and zinc can be found in Yong et al. (2006).

Recently, extensive investigations were performed in the Port Jackson estuary in southeastern Australia, near Sydney, due to the 2000 Olympic Games (Birch and Taylor, 1999). Eight metals were measured in more than 1700 surface sediment samples in the 30-km estuary, river tributaries, harbor annexes, and canals. Copper, lead, and zinc, in particular, were found upstream where there were extensive industrial and commercial activities. Thunderstorms and flooding transported the metals downstream. Total levels of copper, lead, and zinc in the estuarine sediment corresponded to 1,900, 3,500, and 7,300 tonnes, respectively, due to many decades of industrial discharges. Aquatic flora and fauna were affected by the sediments. In the late 1990s, a program for reduction of waste discharges was initiated. However, remediation of the sediments will be complex.

Organic chemical pollutants originate from chemical-producing industries such as refineries, the spillage and leakage of various chemicals such as petroleum products, the use of various products such as paints, greases, oils, pesticides, etc. A common way to group the contaminants is as hydrocarbons, which can be divided into monocyclic, polycyclic hydrocarbons, alkanes, alkenes, etc., or as organohalides, which contain halides such as chlorine. Polycyclic chlorinated biphenyls (PCBs) and trichloroethylene (TCE) are examples of the latter. Organic compounds may also contain oxygen or nitrogen atoms such as methanol or trinitrotoluene (TNT).

The aquatic environment is a resource that must be protected and maintained in a healthy state. When the health of plants and animals that are part of the food chain is impacted, then there is a risk as well to human health. Water is of primary importance for all forms of life. Surface water and groundwater are the primary sources of drinking water. Human activity has numerous influences on the hydrologic cycle. The main processes in the cycle include evaporation and transpiration, condensation, precipitation, infiltration, and runoff.

Humans have significantly altered natural runoff and infiltration patterns and the balance between these two processes. Construction of impermeable surfaces such as roads, highways, and parking lots in urban areas create impermeable surfaces that increase runoff and decrease infiltration. The runoffs subsequently are sent to storm drains or other drainage systems, reducing aquifer levels. Soil compaction during agricultural processes will also increase runoff rates. The transport of contaminants including pesticides, herbicides, insecticides, animal wastes, etc. is increased via runoff, which often reaches surface waters (ponds, lakes, rivers, etc.). Managed runoffs are channeled via sewers and drains and can be discharged with or without treatment. The waters often contain suspended solids that will ultimately become sediments. The dissolved pollutants also may concentrate on the already present sediments. Untreated discharge can reduce water quality substantially. Pollutant source elimination or mitigation of the pollution needs to be practiced, in addition to water treatment.

Water is a highly precious resource. Less than 5% is nonsaline (Yong, 2001), while only 0.2% and 0.3% is found in lakes and rivers, respectively. In addition, more than half to the world's animal and plant species live in the water. Thus protection of water quality is highly important. Decreased water quality decreases the water quantity available, particularly where the need is urgent. In developing countries water use is increasing. Rapid industrialization and urbanization leads to poor water of insufficient quantities. Water management practices need substantial improvement to protect ecosystems and public health. Monitoring of river, lake water, and sediment is not frequently conducted, and therefore the locations of pollutant sources, intensity, and impact are difficult to determine. Only a limited number of parameters such as microbial counts are determined.

Agriculture uses large quantities of water. Water use per crop grown needs to be optimized. Pollutant sources include insecticides, pesticides, fungicides, and fertilizers. Herbicides and pesticides are persistent and can accumulate in animal tissues. Nutrients such as nitrates from runoff of animal wastes from pigs or poultry can severely impact water quality of lakes and rivers (Yamaska River in Quebec, Canada, for example). Detergents are other sources of nutrients. Accumulation of the nutrients can lead to eutrophication and subsequent decreases in water color, taste, and odor. Intensive farming practices have led to increases of phosphorus levels in the lakes. A lack of nutrient treatment processes for wastewater has also contributed. Lake water eutrophication is thus becoming an extensive problem. Elevated nutrients are currently found in many surface waters, and thus even if the input of nutrients is totally eliminated, recovery may take up to 10 years due to slow flushing rates (WHO, 1999).

Elevated levels of nitrogen and phosphorus increase the activity of phytoplankton, macrophytes, and other algal groups. Cyanobacteria, which can fix nitrogen, then may replace phytoplankton, altering the benthic community and other species in the ecosystem. Oxygen becomes depleted, destroying flora and fauna, in the water and at the bottom of the water column. Carbon accumulation occurs, followed by asphyxia and mortality of biota. Nitrogen ingestion by humans can lead to blue-baby syndrome in infants in particular. Sanitary risks can also increase due to ingestion of nitrate-containing water.

Worldwide consumption of fertilizers has increased from 14 to 140 million tons from 1940 to 1999 (Chamely, 2003). One-tenth of these fertilizers contain phosphorus, and one-half contain nitrogen. Farming waste, excrement, inadequately maintained septic tanks, and detergents are other contributing factors. Nitrates are easily leached from the soil because both nitrates and the clay particles in the soil are negatively charged.

Poor agricultural practices and drainage of the fertilizers increases the nitrate contents in many of the European rivers such as the Meuse (4 mg/L), Rhine (3 mg/L), Loire and Po (2 mg/L), and Rhone (1.5 mg/L) (Chamely, 2003). North American rivers such as the Mississippi (1 mg/L) and the Saint Lawrence Rivers (0.25 mg/L) tend to be lower. These levels have been increasing since the 1960s. The other site is the des Hurons River, which is a tributary of the Richelieu River that joins the river on the eastern bank of the Chambly Lake, located east of Montréal, Quebec, and extends 35 to 40 km northeast of the Chambly Lake. The area of the des Hurons River is an intensive farming one with corn and wild plants. The river receives high loads of soil and nutrients due to the agricultural activities in the area. The average of suspended solids (SS) concentration in the des Hurons River in 2007 and 2008 varied from 5.6 mg/L to 134.0 mg/L, and that of chemical oxygen demand (COD) and total phosphorus (T-P) varied from 9.0 mg/L to 26.2 mg/L and 0.05 mg/L to 0.43 mg/L, respectively (Inoue et al., 2009). In Europe, phosphate discharges have been controlled since the ban on phosphate-containing detergents in 1985. Levels increased from 10 to 90 µg/L in Lake Geneva from the 1960s to the 1970s but have decreased to 50 µg/L in the 1990s (Chamley, 2003).

The mining industry discharges their wastes into storage dumps, holding ponds, tailings ponds, and other systems. They can leak, or the structures can fail, allowing discharges into surface water, thus impacting the sediments. Heavy metals in particular are the most common pollutants from mining activities. Other industrial activities contribute due to the increased need for goods due to population growth.

Developed countries exhibit high heavy metal contents in suspended solids and sediments where there has been intense industrial activity. In the Rhine River, concentration of cobalt, copper, and cadmium increased from 1900 to the 1970s. Upon realization of the pollution, implementation of new legislation and modification of industrial processes occurred, thus decreasing heavy metal and other elements in the water and sediments during the period of 1975 to 1985. Concentrations of mercury, lead, zinc, copper, and cadmium increased in the Seine River in France (Meybeck, 2001) from upstream to downstream.

Rivers contribute significantly to the collection and distribution of contaminants. Farm, industrial, and urban wastes end up in the rivers. The rivers then carry the suspended solids to coastal areas. Most artificial sedimentary reservoirs for contaminant and particle trapping are found in Europe, followed by North America, Africa, and Australia.

Atmospheric inputs can also be significant. The wind can carry many pollutants that then fall into water bodies. Nuclear testing and accidents such as the Chernobyl nuclear plant explosion have contributed fallout to nearby and not so nearby regions. Acid rain fallout has also been increasing since the late twentieth century. Estimates are difficult to obtain due to the long-term monitoring requirements and numerical simulations required.

Solid or liquid residues have been dumped for many years into the marine environment. Little monitoring was done prior to the 1970s. Dumped materials included building and construction wastes, industrial, farm, and domestic wastes, chemical and radioactive products, and military products (devices, weapons, and explosives). The dumping has been mainly in deep sea areas greater than 1000 m, although dredging materials can be in more shallow zones closer to the shore.

Oil spills are well-known environmental risks. They have led to serious pollution problems in the Gulf of Mexico, Alaska, Nova Scotia, and in many regions in the English Channel and North Atlantic coast. Less well known is that any other chemicals such as acids, ammonia, heavy metal, fertilizers, pesticides, and other corrosive materials are also transported, and thus spills along the coasts can impact the environment. Long-term dispersion and fate of these chemicals in a marine environment requires better understanding.

Other modes of hydrocarbon transport can be equally or more important than oil spills. Owen et al. (1998) estimated that submarine oil field seepage accounts for 15% of marine pollution, which is three times the amount from oil spills (5%). Other sources include river runoff (41%), tanker dumping or washing (15%), industrial and municipal discharge (11%), coastal refineries and offshore exploration (6.5%), and atmospheric sources (4%). It is likely that offshore exploration will increase due to rising oil prices, and hence the incidences will also become more frequent.

Other contaminants include oil and grease, pesticides, insecticides, and microbial agents. In Lake Geneva, highest levels were found in the period from 1960 to 1975 of the hydrocarbons (poly- and hexachlorobenzene) and DDT insecticides (dichlorodiphenyltrichloroethane) and the breakdown products. The same trends have been seen in the sediments of the Great Lakes of North America (Chamley, 2003).

However, in developing countries, significant pollution problems are occurring. Pathogens are a major problem. The control of pollution is seen as costly and not a priority in areas where lack of food is a substantial problem. Many wastewater plants also do not disinfect their effluents before discharge. Pathogens are known to concentrate in the sediments.

In densely populated areas, reduction of pollutant discharges is the key. Legislative standards have been applied and thus are reducing emissions. The legislation must be implemented and monitored to ensure compliance. However, due to economic pressures on many governments, monitoring is not being strictly carried out. This leaves individuals and enterprises with the responsibility of limiting environmental damage.

Treatment of discharges to reduce toxicity and minimization of water through reuse reduces the entry of toxic substances and suspended solids in to the environment.

The hydrosphere refers to all the forms of water on Earth (i.e., oceans, rivers, lakes, ponds, wetlands, estuaries, inlets, aquifers, groundwater, coastal waters, snow, ice, etc., as seen in Figure 1.1). The geoenvironment includes all the receiving waters contained within the terra firma in the hydrosphere. This excludes oceans and seas, but includes rivers, lakes, ponds, inlets, wetlands, estuaries, coastal marine waters, groundwater, and aquifers. The marine environment in the geoenvironment is included based on the discharge of pollutants in the coastal regions via runoffs on land and polluted waters from rivers or streams.

Microorganisms from agricultural, septic, and sewage discharges are another type of pollutant. They can contribute to the turbidity, odor, and increased oxygen demand in the water. Drinking water contaminated with organisms such as *Escherichia coli* can lead to severe gastrointestinal diseases and possible death. In a small town 200 km north of Toronto, Ontario, Canada (Walkerton), more than 2300 people became ill, and seven died as the result of drinking water from a well contaminated by surface runoff of manure.

1.4 MANAGEMENT OF CONTAMINATED SEDIMENTS

Dredging of sediments is extensively used for maintenance of rivers, harbors, canals, and other areas to ensure boat navigation. For example, in France more than 19 Mm^3 of sediments are dredged to maintain the Seine, Garonne, and Loire estuaries (Chamely, 2003). This activity increases the levels of suspended matter in the water which is subject to transport. In addition, dredged sediments which can contain high levels of contaminants are either landfilled or ocean disposed. Metals, including arsenic, cadmium, copper, mercury, nickel, and lead, PCBs, polycyclic aromatic hydrocarbons (PAHs), pharmaceuticals, and bacterial and viral contaminants are often found in the harbor sediments.

Land disposal is similar to the disposal of other wastes. Incineration, confinement, controlled dumping, and chemical stabilization/solidification are some of the processes employed. Transport of the sediments over long distances may also be required. There is also the potential for return of the sediments to the water due to runoff or leaching of the contaminants. Dredging is often delayed due to management problems, but this can lead to further risks. Ocean disposal can lead to the return of the contaminants to the shore if the currents transport them. Often sediment dumping at sea is at shallow depths near the coast zone to reduce cost. Harbor sediments, in particular, can be contaminated. Recently, Sector 103 of the Port of Montreal was dredged to remove the contamination from heavy metals and hydrocarbons. In a Great Lakes harbor, navigational dredging has not been performed since 1972, because there are no economically and environmentally feasible ways to manage the dredged sediment (USACE, 1995). Ships cannot enter the harbor easily, and loading/ unloading is becoming problematic. All of these problems increased transportation costs in the Indiana Harbor Ship Canal and decreased shipping capacity by 15%.

Reduction of pollutant release at the source is required to prevent accumulation of the contaminants in sediments from both point and nonpoint sources. Prevention and source control programs are required to ensure this. For example, industrial plants must have adequate treatment and storage systems. Mining is a major source of heavy metal discharges. Domestic and sanitary sources provide organic inputs into the waterways. Agricultural fertilizers should not be overused and should be applied as needed by the plants. Some measures for reduction have been discussed previously (Yong et al., 2006).

Water is a major transporter of the contaminants from lakes and rivers into the oceans. In addition, contaminants can be trapped in the sediments of artificial lakes and waterways where natural discharges are not possible. Strict regulations and sediment quality monitoring are required. Inventories of sediment quality are needed. Many have not been updated for many years. As previously discussed, reduction of the inputs can significantly improve sediment quality. One of the key areas of concern for accumulation of contaminants in the food chain has been the Great Lakes area. Many years of industrial and municipal discharges have occurred, but little attention was paid to the state of the bottom sediments until the 1980s. The EPA Great Lakes National Program Office (GLNPO) has indicated that contaminated sediments are the most significant source of contaminants for the food chain in the Great Lakes rivers and harbors. There are 42 Areas of Concern (AOC), and as a result more than 1.8 million cubic m^3 have been removed from the Basin from 1997 to 2002 (http://www.epa.gov/glnpo/glindicators/sediments/remediateb.html). In 2004, more than 3,221 advisories for the limitation of fish consumption were issued. More than 35% of the total lake area, 24% of the total river lengths, and 100% of the Great Lakes and connecting waters were covered, mainly because of the contamination of the sediment (USEPA, 2005). Navigational dredging in harbors and ports is often not completed due to the cost and concern for water quality and sediment disposal issues. The Superfund program started to take action regarding about 140 contaminated sediment sites. The most frequently found contaminants were PCBs (44%) and metals (39%), followed by PAHs (24%), mercury (15%), pesticides (12%), and a mixture of others (14%).

Much progress has been made in developed countries with regards to recycling and reduction of pollutants at the source. Tools for the evaluation and characterization of contaminants in the sediments will be discussed in a later chapter (Chapter 3). However, in developing countries the challenges are substantial.

In the ocean, discharges at sea must be reduced. This can allow the natural purification processes to be exploited. Not only are discharges directly into the sea problematic but incineration of wastes (vinyl chlorides, PCBs) is also practiced (Salomons et al., 1988). Dumping has been regulated since the 1970s and consists of building, construction, and demolition products, industrial, farm, and domestic wastes, and chemical/radioactive wastes. Most authorized disposal areas in deep seas of more than 1000 m in depth are located in the Atlantic Ocean near England, Canada, and the United States. The dispersion of smoke and particles into the atmosphere can potentially occur and lead to fallouts into the oceans. Mercury can accumulate in ocean sediments due to this mechanism. River discharges can carry land particulates to the sea

Erosion and dredging can disperse contaminated sediments into previously uncontaminated areas. The effects of dredging need to be minimized and beneficial uses of dredged material promoted as much as possible. Eutrophication from elevated nutrient inputs and bioaccumulation in the food chain disrupt the hydrosphere, biosphere, and lithosphere. Remediation management tools as an alternative to dredging are also required. Monitoring is required to ensure that risk management objectives are achieved and that source control and prevention are carried out adequately.

1.5   NATURAL MITIGATION PROCESSES

More recently, the assimilative capacity of sediment materials has been exploited as a means to attenuate the contaminants in the contaminated sediments. The term natural recovery (NR) has been used to identify the results of contamination attenuation in the sediments through natural processes. The processes involved are in almost all respects similar to those available in the corresponding natural attenuation (NA) treatment processes used in the solid land environment and have been described in Yong and Mulligan (2004). The primary processes involved in NR fall under the category of bioremediation or biotransformation. These are complex processes that are not only conditioned by the natural microbial communities and metabolic processes, but also by the nature of the organic compounds and the other sediment components.

Natural purification processes include sorption, precipitation, biodegradation, dilution, and dispersion (Figure 1.5). These processes are known as natural attenuation or, in the case of sediments, natural recovery. These will be discussed in more detail in Chapter 5. Contaminants can accumulate for decades due to sedimentation in the bottom of lakes and rivers. The risk of contamination of the water due to propeller and boat movement is increased. During floods, sediment erosion is enhanced. Fluctuating pH conditions also can release poorly bound or unstable fractions of oxides and organic complexes.

Natural processes for the reduction of the amounts of contaminants in the sediment have been utilized as a means of purification. Dilution and bacterial activity are the main processes. However, due to excessive pollutant inputs in areas such as the Baltic and Mediterranean Seas (Chamely, 2003), the natural properties of recovery have diminished. In the marine environment in particular, the potential requires much more understanding to assist in long-term management measures. The DDT group of insecticides has been discharged off the California coast since 1970 (Zeng and Venkatesan, 1999). The sediments have shown decreasing levels due to biodegradation and dilution. Additional particle trapping and diffusion outside of the discharge zone may also have occurred.

The Great Lakes of North America include Superior, Huron, Michigan, Erie, and Ontario. They are the largest fresh water system in the world (USEPA, 1995). More

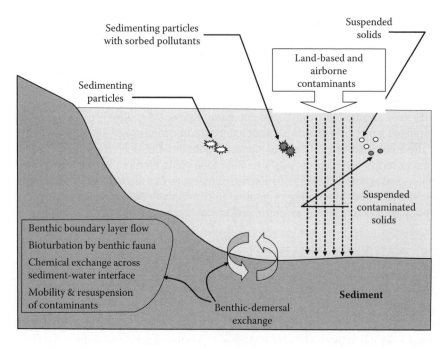

FIGURE 1.5 Processes contributing to contamination and retention of contaminants in the surface sediment layer.

than 33 million people live near the water. The water is used for consumption, transportation, agriculture, power, recreation, extraction of natural resources, and other uses. Although at one time 180 fish species lived in the Great Lakes, the number has decreased substantially due to pollution and loss of habitat. Contaminated sediments have led to commercial and recreational fishing advisories. Fish tumors and deformities, degradation of phyto- and zooplankton, eutrophication, and the growth of undesired plants have increased, and degradation of aesthetics has been noted. Although impacts must be reduced, sediment remediation has been slow due to a lack of information regarding the sources and extent of the problem and a lack of cost-effective remediation technologies, funding limitations, and political problems. Natural recovery is an attractive solution at many sites in the Great Lakes. Various case studies will be examined in Chapter 5.

1.6   BIOACCUMULATION OF CONTAMINANTS

As sediments are a reservoir for contaminants, the fish and benthic organisms that live within them can accumulate toxic compounds. The levels can bioaccumulate up the food chain to birds, fish, and other animals to toxic levels. Neurological, developmental, and reproductive problems may manifest. In the United States, the EPA reported that more than 2100 state advisories were issued due to health hazards from consuming fish (USEPA, 1998). Ninety-six watersheds were identified as "areas of probable concern." Sediment toxicity tests are now used to evaluate sediment contamination.

The EPA has also proposed that pollution prevention measures should include development of guidelines for new chemicals based on bioavailability and partitioning to sediments. A list of some of the effects of various chemicals is shown in Table 1.1.

Sediment Quality guidelines of the Canadian Ministry of the Environment (CCME) are based on the chemical concentration in the sediment that causes an effect on aquatic species (CCME 1999). Two reference values are established, the threshold effect level (TEL) and the probable effect level (PEL). Recently in Quebec, three additional reference values were added, the rare effect concentration (REL), the occasional effect level (OEL) and the frequent effect level (FEL) (Environment Canada and the Ministère de Développement durable, de l'Environnement et des Parcs du Québec, 2008). The two latter effect levels are to be used for management of dredged sediment disposal and remediation decisions. A full list of the assessment quality criteria levels for a wide variety of chemicals is presented in Appendix A for fresh and marine sediments. The guidelines have limitations, including the lack of incorporation of bioaccumulation (accumulation in biological tissues) and biomagnification (accumulation as the concentration goes up the food chain), and the absence of consideration of other effects such as elevated suspended solids levels and loss of habitat. The effect on specific species is not considered, in addition to additive, synergistic, or antagonistic effects. PCBs, pesticides, and methyl mercury are examples of contaminants that both bioaccumulate and biomagnify.

Contamination and its linkage to society is complex. Economic development and increasing population put pressure on the environment. How technological develop-

TABLE 1.1
Chronic Effects of Some Hazardous Wastes

Waste Type	Effect
Pesticides	Nervous system, liver, kidney effects, possible carcinogen, mutagen, teratogen
Herbicides (2-4-D* and others)	Nervous system, liver, kidney effects, possible carcinogen, mutagen, teratogen
Polychlorinated biphenyls	Potential carcinogen, teratogen
Halogenated organics	Carcinogenic and mutagenic risk
Nonhalogenated volatile organics	Potential carcinogen and mutagen
Zn, Cu, Se, Cr, Ni, Pb	Liver and kidney effects, cancer risk
Hg	Nervous and kidney effects, mutagenic and teratogenic risk
Cd	Kidney deficiency, cancer risk
As	Dermal and nervous system toxicity effects, cancer risk
Cyanides	Poisoning
Fecal contaminants	Potential digestive system risks

Source: Adapted from Governor's Office of Appropriate Technology, Toxic Waste Assessment Group, California, 1981, adapted from Chamley, 2003.

* No reportable information available.

ment and exploitation of the natural processes can be used to minimize environmental risk will be a subject of this book.

1.7 SUSTAINABLE SEDIMENT MANAGEMENT PRACTICES

Once sediments are identified as contaminated after an investigation that indicates the potential for risk to human health, fisheries, or the environment, then a remediation methodology must be developed. Strategies for remediation of contaminated sediments must consider the combination of (1) the nature and distribution of the contaminated sediments, (2) determination of the nature, properties, and characteristics of the sediments, (3) development of the necessary remediation treatment technologies that will successfully remove the contaminants from the sediments and minimize risk during and after remediation, and (4) applying the necessary technological evaluation and monitoring to support the decontamination treatment and ensure the sustainability of the remediation.

Present remediation procedures tend to either remove the contaminated sediments or employ in situ methods to manage contaminated surface sediments. To a large extent, these methods effectively reduce the bioavailability and transfer of contaminants into the water column. In situ chemical or biological treatment and natural processes can be used. Treatment options of dredged materials should also be considered, particularly to ensure beneficial uses (USACE/USEPA, 2004). This will contribute to the reduction of the use of nonrenewable geological resources. In the 1990s, according to Forstner and Apitz (2007), removal was the main approach utilized in North America and Europe. However, due to the substantial costs for removal of large volumes and the risks to the environment, in situ management approaches are becoming more acceptable.

The most common techniques include:

- Environmental dredging following by drying and sediment handling
- Sediment treatment of dredged materials by physical, chemical, and biological processes
- Containment in contained disposal facilities (CDFs), contained aquatic disposal (CAD), and landfills
- In situ capping
- Monitored Natural Recovery (MNR)
- In situ treatments by chemical or biological processes

Selection of the most appropriate method is difficult and has been the subject of much discussion. This book will discuss selection criteria and aspects to be considered during the evaluation of the remediation technology. The process of the evaluation will involve the following steps: characterization and assessment of the problem, source control implementation, site and sediment characterization, comparison and assessment of the remediation alternative, selection of the remediation, and determination of the monitoring and management methodology. Knowledge of the nature and composition of contaminated sediment is required to avoid resuspension and remobilization of contaminants. The information obtained will also allow one to determine the best or most effective means for treatment for remediation of the contaminated sediment—consistent with cost-effective considerations. Limitations of each alternative will be addressed. Each step of this process will be discussed in this book. Mixtures of heavy metals, hydrocarbons, and chlorohydrocarbons pose substantial challenges. Regulatory standards and criteria must be met.

Consultation with the public should be done at all phases. This will enable concerns to be identified and addressed early. Some concerns include (USEPA, 2005):

- Human health impacts
- Ecological impacts
- Loss of recreational activities
- Loss of fisheries, property values, development opportunities, tourism
- Identification of all contaminants and their sources
- Loss of commercial navigation
- Loss of traditional cultural aspects by native tribes

Sustainability is an additional element to be considered in an effort to work toward meeting the goals of sustainability. Resource conservation and management and preservation of diversity are included. If not, the capability of the aquatic geoenvironment to provide the basis of life support will be diminished. To be sustainable, the sediments would need to remain harmless over a long period of time. Ultimately, the habitat should be restored to enable species preservation and biodiversity regeneration. Sustainability, here, refers to the ability of the system to maintain or preserve the initial condition, state, or level before contamination. Sustainability of remediated sediments refers to the ability of the remediated sediments to be preserved in the remediated condition. The key to a sustainability assessment is to minimize and/or eliminate health threats to humans.

Revitalization of land and water areas is another key aspect to be considered in evaluating the sustainability aspects. The use of waterfront properties, harbors, and water bodies can be substantially enhanced and revitalized by sediment decontamination projects. The various aspects of the Lachine Canal project in the Montreal area will be discussed in Chapter 9. Land use plans should be reviewed, and land owners and planning and development agencies should be consulted.

For a remediated sediment treatment to become sustainable, (a) the sediment must not require retreatment to maintain its remediated state, and (b) it must reestablish its original uncontaminated benthic ecosystem. Retreatment of contaminated remediated sediments is not desirable for many reasons and needs to be avoided.

Information on the sources of contaminants provides the nature and composition of the contaminants. These can be numerous and difficult to identify. They may also lead to diffuse contamination over large areas. Knowledge of the sources of contaminants will provide the means for developing regulations and strategies for managing or controlling the discharge of contaminants that would eventually find their way into the receiving waters and impact the sustainability of any remedial action.

The various strategies for remediation of contaminated sediments provide for different results concerning how the contaminants in the sediments are neutralized or eliminated. Some of the remediation techniques available are listed in Figure 1.6.

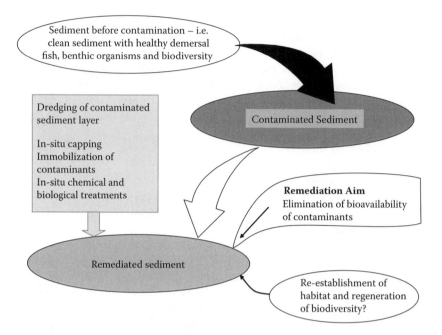

FIGURE 1.6   Alternatives for remediation of contaminated sediments.

The nature of the remediated sediment will have a direct influence on the strategies and capabilities for sustainability of the remediated sediment to be achieved. These techniques will be examined in later chapters. The requirements for remediated sediment sustainability are controlled by the information from the short- and long-term human health risks, regulatory attitudes and goals, economics, and site specifics. Decision frameworks must be based on a good scientific knowledge of the site.

Figure 1.7 shows that the primary source for resuspension and remobilization of contaminants in the sediment is the top portion of the contaminated surface sediment layer. Bioturbation and benthic boundary layer flow, including tidal exchange, will most likely affect only about the top 30 cm of the surface sediment layer. This figure shows some of the difficulties of remediating sediment sites. These natural forces influence contaminant transport.

1.8   CONCLUDING REMARKS

Proper management of the aquatic geoenvironment is needed to protect future generations, but is highly complex. Water quality must not be degraded so that it cannot be consumed without risk to the health. The same follows for all resources obtained from the water. Sediment as a natural resource must not be depleted through quality degradation. Technologies for environmental management for remediation and impact avoidance would reduce the degradation of sediment quality and will be examined thoroughly in the following chapters. Protocols and procedures to monitor and manage changes in the environment will also be required and will be discussed in this book.

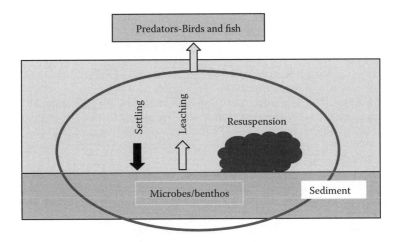

FIGURE 1.7   Interactions of abiotic and biotic elements.

Obtaining sustainable remediated sediment requires (a) source control of contaminants entering the ecosystem, (b) natural processes within the surface sediment layer that maintains the remediated state of the sediment, and (c) restoration of habitat and reestablishment of biodiversity. Human intervention in providing the necessary elements for restoration of habitat and reestablishment of biodiversity, after or during remediation of the contaminated sediment, will provide for sustainable remediated sediment. However, it must be done in a cost-effective manner.

REFERENCES

Birch, G. and Taylor, S. 1999. Source of heavy metals in sediment of the Port Jackson estuary, Australia. *Sci. Total Environ.* 227: 123–138.

Brils, J.M. 2003. The SedNet Strategy Paper: The opinion of SedNet on environmentally, socially and economically viable sediment management. SedNet. http://www.SedNet.org.

CCME 1999. Canadian Sediment Quality Guidelines for the Protection of Aquatic Life. Canadian Council of Ministers of the Environment. CCME EPC-98E http://www.ccme.ca/assets/pdf/sedqg_protocol.pdf.

Chamley, H. 2003. *Geosciences, Environment and Man.* Elsevier, Amsterdam.

Environment Canada and Ministère du Développement durable, de l'Environnement et des Parcs du Québec. 2008. Criteria for the Assessment of Sediment Quality in Quebec and Application Frameworks: Prevention, Dredging and Remediation.

Förstner, U. and Apitz, S.E. 2007. State of the art in the USA. Sediment remediation: U.S. focus on capping and monitored natural recovery. *J. Soil Sediments,* 7(6): 351–358.

Inoue, T., Mulligan, C.N., Zadeh, E.M. and Fukue, M. 2009. Effect of contaminated suspended solids on water and sediment qualities and their treatment. *J. ASTM International.* 6(3), pages 1–11, Paper ID JAI102185.

Meybeck, M. 2001. Transport et qualité des sediments fluviaux: Variabilités temporelle et spatiale, enjeux de gestions. Publication de la Société Hydrotechnique de France. 166th sess, 11–27.

Owen, O.S., Chiras, D.D. and Reganold, J.P. 1998. *Natural Resource Conservation.* 7th ed. Prentice Hall, 594p.

Salomons, W., Bayne, B., Duursma, E.K. and Forstner, U. 1988. *Pollution of the North Sea: An Assessment.* Springer, Berlin.

Trevors, J.T. 2003. Editorial: biodiversity and environmental pollution. *Water, Air Soil Pollut.* 150:1–2.

USACE 1995. Ecosystem Restoration in the Civil Works Program. ER-1105-2-210. USACE (U.S. Army Corps of Engineers), Washington, DC.

USACE/USEPA 2004. Evaluating environmental effects of dredged material management alternatives—A technical framework. EPA 842-B-92-008, U.S. Army Corps of Engineers and U.S. Environmental Protection Agency. Washington, DC.

USEPA 1995. Water quality guidance for the Great Lakes system. USEPA. Federal Register 60: 15366-153425.

USEPA 1998. EPA's Contaminated Sediment Management Strategy. EPA-823-R-98-001, United States Environmental Protection Agency, Office of Water, Washington, DC.

USEPA 2002. *Contaminated Sediment Remediation Guidance for Hazardous Waste Sites.* United States Environmental Protection Agency, Office of Solid Waste and Emergency Response, Washington, DC. Report for OSWER 9355.0-85.

USEPA 2005. *Contaminated Sediment Remediation Guidance for Hazardous Waste Sites.* Environmental Protection Agency, Office of Solid Wastes and Emergency Response, EPA-540-R-05-012, OSWER 9355.0-85, Washington, DC.

Varanasi, U., Casillas, E., Arkoosh, M.R., Hom, T., Misitano, D.A., Brown, D.W., Chan, S.-L., Collier, T.K., McCain, B.B., and Stein, J.E. 1993. Contaminant exposure and associated biological effects in juvenile Chinook salmon (*Oncorrhyncus tshawytscha*) from urban and nonurban estuaries of Puget Sound, Seattle, WA. National Oceanic and Atmospheric Administration (NOAA) National Marine Fisheries Service, NMFS NWFSC-8.

WHO 1999. *Toxic Cyanobacteria in Water: A Guide to Their Public Health Consequences, Monitoring.* Chorus, I. and Bartram, J. (Eds.). W& FN Spon, London and New York.

Yong, R.N. 2001. *Geoenvironmental Engineering: Contaminated Soils, Pollutant Fate and Mitigation.* CRC Press, Boca Raton, FL.

Yong, R.N. and Mulligan, C.N. 2004. *Natural Attenuation of Contaminants in Soils.* CRC Press, Boca Raton, FL.

Yong, R.N., Mulligan, C.N., and Fukue, M. 2006. *Geoenvironmental Sustainability.* CRC Press, Boca Raton, FL. 387 pp.

Zeng, E.Y. and Venkatesan, M.I. 1999. Dispersion of sediment DDTs in the coastal ocean off southern California. *Sci. Total Environ.* 229: 195–208.

2 Introduction to Sediments

2.1 INTRODUCTION

Parent rock can be broken down by physical and chemical weathering. Weathered rocks, such as coarse grains and clay minerals, are transported by the flow of water and are deposited in rivers, lakes, estuaries, and sea areas. These materials form the sediments in water. On the other hand, smaller particles and materials dispersed from volcanoes and transported by the wind are called aeolian soil upon deposit.

Soil organic matter degraded by microbial activities is also discharged into rivers, lakes, estuaries, and sea areas with inorganic particles. Furthermore, a variety of substances are discharged through human activities. These are fed into water through various channels. Thus, sediments are in contact with inorganic, organic, and other human-discharged materials, through the influence of the pore water. Therefore, the properties of the pore water are an important factor regarding the quality of the sediments. Since the industrial revolution, around 1750, a variety of production and consumption activities by humans have also created problems associated with waste materials, because some of the wastes which have been discharged and have accumulated are hazardous or toxic (Cappuyns et al., 2006; Fabris et al., 1999; Fukue et al., 2007).

Fine particles discharged from land that are easily suspended in water are called "suspended solids." They can agglomerate in the water and start settling. Generally, these particles consist of many fine materials such as clay minerals, organic matter (including plankton), oxides and hydroxides, etc. They often adsorb nutrients and hazardous substances, in addition to bacteria and viruses. Figure 2.1 shows an example of settling particles in a brackish lake.

There are basically two types of problems related to sediments: eutrophication with nutrients and contamination with hazardous and toxic substances. Basically, the adsorption and desorption of contaminants and the degradation of organic particles can become problematic. Eutrophication and contamination cannot be treated in the same way, because contaminants are hazardous, whereas nutrients are not.

Sediments are often called "mud," "sand," and "gravel," depending on the nature of the deposited materials. In fact, this classification has been used for anchoring ships, and the terms are indicated in the chart. Although these names are mostly due to the size of sediment particles, they are only generic names. For example, when sediments are organic rich, or smaller particles, they are often called "mud." However, there are detailed classification methods for sediments for scientific and engineering purposes.

Sediments are often called marine soils or lake soils. They are also called marine-deposited soils or lake-deposited soils. These are all contained in the categories of

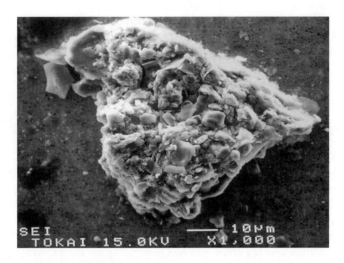

FIGURE 2.1 Settling particles in a brackish lake.

sediments. Therefore, sediments are defined as any solid deposited at the bottom of a ditch, river, lake, or sea. Accordingly, they are called ditch sediments, river sediments, lake sediments, or marine sediments, respectively. They can also be separated into fresh and marine sediments. The intermediate sediments are sometimes called brackish sediments. From an environmental point of view, sediments in fresh water are distinguished from sediments in seawater because of the different associated food chains and biological concentrations. For example, Canadian guidelines provide more severe values for the guidelines for marine sediments than for freshwater sediments.

In coastal regions, the origin of sediments is mostly the land. However, a portion of the sediments is produced in water. They are of plankton, authigenic minerals, crusts, aquatic plants, or other organic origins. Some of these products are buried with deposits from land, and therefore, the ratio of marine produced to discharged materials from land is lower near the coasts, but will increase with distance from the coast. At ocean bottoms far from the coast, the effect of land materials is small, and thus the sediments consist mainly of products formed under marine conditions. A typical example is ooze (Wetzel, 1989). Wetzel (1989) investigated the consolidation of ooze in deep ocean. Diagenesis that occurred there formed chalk and sedimentary limestone layers.

In general, larger discharged solid particles are transported near the shore and bottom by water currents and waves, whereas finer particles will disperse further. They finally settle on the bottom but can move again, depending on the water action. In this sense, surface sediments are more active than the underlying sediments. Therefore, erosion is dependent on the balance of the settlement and movement of the sediments. Thus, the reduction in the discharge can cause coastal erosion. In many cases, the control of soil discharge in mountainous regions has caused coastal erosion.

FIGURE 2.2 Pyrites formed in marine sediments.

New minerals can form in sediments due to the change in reduction-oxidation conditions. Pyrite is a typical product formed under marine reduced conditions. If pyrite is found in sediments (Figure 2.2), it means that there is a lack of dissolved oxygen and a low reduction-oxidation (redox) potential.

2.2 DEFINITION OF SEDIMENTS

In this book, sediments are defined as solids that have settled and deposited at the bottom of the water column. They contain liquid and gas phases, which is similar to the definition of soil in geotechnical engineering. Therefore, sediments consist of sediment particles (solid), pore water (liquid), and gas. The liquid may be fresh water, seawater, wastewater, or a mixture of them. The gas may be air, methane, another type of gas, or a mixture of gases. In this sense, marine-deposited soils and fluvial sediments can be classified as sediments. However, problems with those sediments have often been dealt with in a manner similar to soil and groundwater, because they have required different approaches from sediments under water. Therefore, sediments within groundwater are distinguished from the sediments under surface water in this book.

2.3 TYPES OF SEDIMENTS

There are some classification methods corresponding to the objectives. Therefore, the type of sediment depends on the methods of classification. The determination of the classification of sediments requires some testing and/or analyses. Since the quality of sediments can be obtained by evaluating the properties, the selection of test methods and analyses is important.

Sediments can be primarily classified into three categories (i.e., marine, fresh, and brackish sediments). The term brackish is mainly associated with the water fraction, not the sediment. The word brackish comes from the Middle Dutch root "brak" which means salten or salty. Usually, brackish is used for water having a salt concentration greater than 0.5 up to 30 parts per thousand (ppt), whereas seawater has a concentration between 30 and 50 ppt.

The Baltic Sea is a brackish sea adjoining the North Sea. Initially, prior to the Pleistocene epoch two major river systems met at this location. However, although it was flooded by the North Sea, it still receives a sufficient quantity of fresh water from the adjacent lands that makes the water brackish (http://en.wikipedia.org/wiki/Brackish_water#Brackish_seas_and_lakes).

There are many brackish lakes and rivers that are connected to the sea. Estuaries are where seawater and fresh water mix, and consequently they are under brackish conditions.

2.3.1 Types of Sediments by Components

Sediments are composed of various components, as shown in Figure 2.3. Solids in the sediments include organic and inorganic particles. In general, organic particles have lower specific gravities than that of inorganic particles. If a sediment has a specific gravity lower than 2.5, it usually contains an ignition loss greater than 15%. This is compared to the typical value of 2.65 for primary minerals, such as feldspar and quartz.

Sediments are evaluated according to the organic matter content. If the ignition loss or total organic carbon is high, the sediment is called organic sediment or sediment with organic matter.

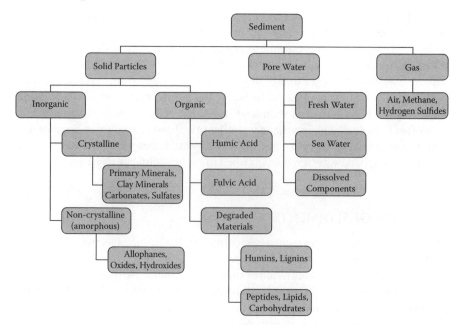

FIGURE 2.3   Various components of sediments.

Pore water can include fresh water, seawater, or other liquids like wastewater. The quality of the pore water is very important from an environmental point of view, because it affects aquatic life. The quality of the pore water can be analyzed with chemical procedures using appropriate sampling techniques.

Pore gas can also be important in characterizing some types of sediments, because anaerobic conditions will produce methane gas and hydrogen sulfide. The existence of CO_2 is also possible. Figure 2.4 shows the solid phase of the sediments for both the inorganic and organic components. The inorganic components can be crystalline or noncrystalline. More detail can be found in Yong and Mulligan (2004), but briefly they can be described as in the sections that follow.

2.3.1.1 Primary Minerals

Primary minerals are derived from the parent rock material through mainly physical weathering processes. Primary minerals include quartz, feldspar, micas, amphiboles, and pyroxenes and are generally found as sand and silt fractions. However, quartz is chemically stable and exists as a small portion of clay-sized fractions.

2.3.1.2 Secondary Minerals

Secondary minerals are formed by physical, chemical, and/or biological weathering processes. These minerals are layer silicates and are known as phyllosilicates. They comprise a major fraction of the clay-sized materials in clays.

Clays and clay soils refer to soils that have particle sizes less than 2 micron effective diameter (draft by ISO). However, some countries like the United States use different definitions for clay fractions (i.e., <5 microns). Note that clay or clay soils and clay minerals are not always the same. Clay or clay soils are defined by size. On the other hand, clay minerals refer specifically to the layer silicates. These are secondary

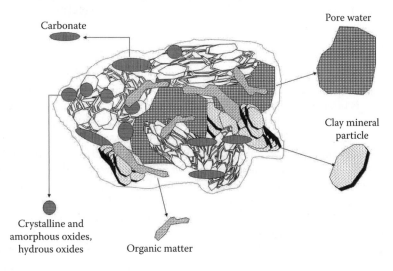

FIGURE 2.4 An idealized typical sediment consisting of various fractions (from Yong et al., 2006).

minerals consisting of oxides of aluminium and silicon with small amounts of metal ions substituted within the crystal structure of the minerals. Due to their size and structure, secondary minerals have large specific surface areas and significant surface charges. The major groups of clay minerals include kaolinites, smectites (montmorillonites, beidellites, and nontronites), illites, chlorites, and vermiculites.

2.3.1.3   Organic Matter

Organic matter originates from vegetation and animal sources and is generally categorized into humic and nonhumic material. Humic materials are those organics that result from the chemical and biological degradation of nonhumic material. Nonhumic material or compounds, on the other hand, are organics that remain undecomposed or are partly degraded. Humic substances are classified as humic acids, fulvic acids, and humins, based on their solubility in acids and bases.

2.3.1.4   Oxides and Hydrous Oxides

Oxides and hydrous oxide minerals includes the oxides, hydroxides, and oxyhydroxides of iron, aluminium, manganese, titanium, and silicon. Common crystalline forms of these minerals include anatase, bohemite, gibbsite, haematite, goethite, and quartz. They are different from layer silicate minerals (secondary minerals) because their surfaces consist of broken bonds with hydroxyl (OH^-) groups of disassociated water molecules. The surfaces exhibit pH-dependent charges.

2.3.1.5   Carbonates and Sulfates

The most common carbonate mineral found in sediments is calcite ($CaCO_3$). Some types of plankton may incorporate calcite into their shell. Some of the other less common forms are magnesite ($MgCO_3$) and dolomite ($CaMg(CO_3)_2$). Gypsum ($CaSO_4 \cdot 2H_2O$) is the most common sulfate mineral found in sediments. Carbonates can retain hazardous materials by the precipitation of heavy metals.

2.3.2   Types of Sediments by Grain Size

The grain size of particles is not their actual size. Sediment particles have a variety of shapes. Therefore, the accurate size of the sediment particles is difficult to define. The grain size is often defined as the equivalent size of a sphere, or as the sieve opening.

Sediments are often classified by the grain size and its distribution. The most commonly used method to determine the grain size of sediments is according to geotechnical engineering techniques (ASTM D422, JIS A 1204). These methods use sieves for coarse particles and a hydrometer for finer particles, as will be described in a later chapter. These techniques require a relatively large amount of sample.

When only a small amount of sample is available, laser diffraction methods can be used (Fukue et al., 2006; Wen et al., 2002). Various counters for grain size analysis are now commercially available for sediment, soil, and powders. Figure 2.5 shows the definition of grain size. ASTM provides the classification as gravel, sand, silt, clay, and colloid in order from large to small particles. The term gravel is used to describe particles larger than 2 mm. Sand consists of mineral particles smaller than gravel, but larger than 0.075 mm. They may be coarse, medium, or fine, depending on the size

ASTM	Gravel		Sand	Silt	Clay	Colloid
AASHTO		Gravel	Sand	Silt	Clay	Colloid

| | 76.2 | 2.00 | 0.075 | 0.005 | 0.001 Size in mm |

ASTM: American Society for Testing and Materials

AASHTO: The American Association of State Highway and Transportation Officials

FIGURE 2.5 Definition of particles by grain size.

of the particles. A silt classification is used for mineral particles ranging in size from 0.005 to 0.075 mm. Clay-size particles are smaller than silt size, without consideration of the mineral content. If the particles are smaller than 0.001 mm, they may be called colloids. Clay minerals are complex hydroaluminum silicates ($Al_2O_3 \cdot nSiO_2 \cdot kH_2O$, where n and k are numerical values of the number of molecules attached).

The sediment is classified by grain size as illustrated, in Figure 2.5, which shows a typical example of the grain size distribution of sediments and its classification according to size. Figure 2.6 shows the grain size distribution curves obtained for a lake core sample. The vertical axis provides the percent of weight of the total particles finer than the grain size in the horizontal axis. Terminologies, such as gravel, sand, silt, and clay are used for a given range of grain sizes. It is important to note that the classification varies slightly according to ASTM and ISO standards. The term "colloid" is sometimes used for particles smaller than 0.001 mm.

In geology, another type of grain size classification is used. The Krumbein *phi* (φ) scale, a modification of the Wentworth scale created by W.C. Krumbein (Krumbein and Sloss, 1963), is a logarithmic scale computed by the equation:

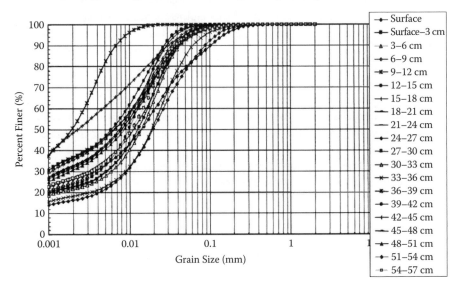

FIGURE 2.6 Examples of grain size distribution curves of sediments.

$$\varphi = -\log_2 D/D_0 \qquad (2.1)$$

where
φ is the Krumbein phi scale, and
D is the diameter of the particle
D_0 is a reference diameter, equal to 1 mm.
This equation can be rearranged to find the diameter using φ:

$$D = D_0 \times 2^{-\varphi} \qquad (2.2)$$

The distribution of grain size is not a parameter, but a state. Therefore, various parameters are defined and determined from the distribution curve and used in engineering practices. For example, D_{10} (mm) is the grain size at which 10% of sample is finer. It is often called an effective grain size. D_{30} and D_{60} are defined as the grain sizes at which 30% and 60% of the sample are finer, respectively. The uniformity coefficient of grain size, Uc' is defined by D_{60}/D_{10}. If Uc' is smaller than 10, it means that the sediment particles are uniform in size. It is noted that the properties of sediments are not evaluated from the Uc' only, because the properties of sediments are also dependent on the particle size and mineral type. Therefore, the maximum and minimum grain sizes are also taken into account to evaluate the properties of sediments.

2.3.3 STRUCTURE OF SEDIMENTS

The structure of sediments is dependent on many factors, such as grain size and shape, mineralogical aspects of particles, the organic matter content, and the ionic strength of the pore water, etc. If the sediments are in a muddy state, the structure may be dependent on the type of suspended solids, since the suspended solids became sediments. In many cases, the structure of surface sediments is described as a stack of flocs, which are aggregates of small particles. The structure is very loose, because macro-pores are formed between the flocs, as shown in Figure 2.7. Coarse particles deposit near the coast, especially near the mouth of rivers, and on beaches where waves and currents are strong enough to wash fine particles away. In general, the further from the coast, the smaller the grain size of the sediments.

The pore volume of the sediments can be evaluated from the water content w, where water content is defined from Figure 2.8 as

$$w = \frac{m_L}{m_s} \times 100 \quad (\%) \qquad (2.3)$$

where m_L and m_s are the masses of the pore water and solids, respectively. The water content is determined by drying the sediment sample at 110°C using an oven. The m_s is the weight of sample after drying, and the m_L is the reduction of the weight by

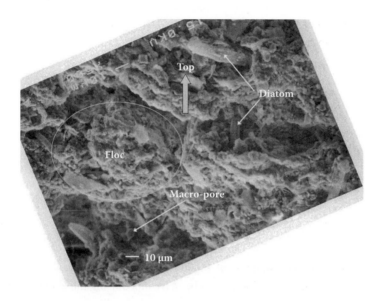

FIGURE 2.7 Microphotograph of surface sediments from a brackish lake.

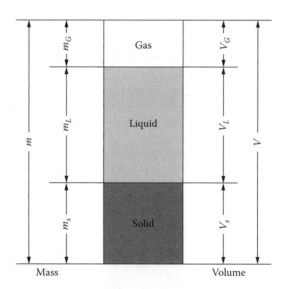

FIGURE 2.8 Three phases of sediments in terms of mass and volume.

drying. The volume of the pore can be expressed by void ratio e or porosity n, which is defined from Figure 2.5 as

$$e = \frac{V_G + V_L}{V_s} \quad (2.4)$$

Since sediments are usually saturated with water, the void ratio is then

$$e = \frac{V_L}{V_s}$$

$$= \frac{w}{100} G_s \quad (2.5)$$

where V_L and V_s are the volumes of liquid and solids, respectively, and G_s is the specific gravity of particles. The G_s can be determined using the standard methods provided by ASTM or other organizations. The relationship between void ratio (e) and porosity (n) is illustrated in Figure 2.9, where

$$n = \frac{e}{1+e} \times 100 \quad (\%) \quad (2.6)$$

The dry density of sediments, ρ_d can be determined (Fukue and Mulligan, 2009) using the parameters obtained as

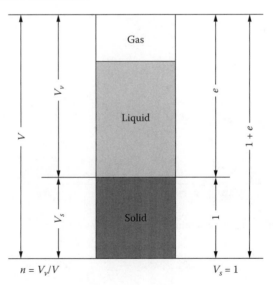

FIGURE 2.9   Void ratio and porosity.

$$\rho_d = \frac{m_s}{V} = \frac{G_s\rho_L}{1+e} \tag{2.7}$$

where ρ_L is the density of liquid. The bulk density, ρ_t is given by

$$\rho_t = \frac{m_L + m_s}{V} = \rho_d\left(1 + \frac{w}{100}\right) \tag{2.8}$$

The bulk density is determined by measuring the weight and volume of the sample. The properties of various sediments are shown in Table 2.1. The specific gravity of particles varies slightly. On the other hand, the water content varies over a wide range, from about 30% in sandy sediments to more than 400% in fine sediments. The void ratio and porosity are related to water content, because of the fully saturated state.

Organic matter can influence the structure and other chemical properties of sediments. Ignition loss is a good indication of the organic content in sediments, which is expressed by

$$I_L = \frac{\text{reduction of weight after burning}}{\text{initial dry weight of the sample}} \times 100 \quad (\%) \tag{2.9}$$

Various burning temperatures have been recommended by different organizations. Generally, a range between 375°C and 850°C has been used for the test of ignition loss, depending on the soil type and components to be measured.

The water content of the muddy sediments obtained from a sea port is strongly influenced by the ignition loss (750±50°C) as shown in Figure 2.10. The higher the ignition loss, the higher the water content. This is not because of the portion of organic matter, but the structure is dependent on the organic matter. This trend can be seen in the sediment profile in Figure 2.11, which shows the relationship between water content and ignition loss in a core sample obtained from a brackish lake. In this case, the consolidation effect on the water content decreases the effect of ignition loss.

TABLE 2.1

Properties of Lake, Coastal, and Bay Sediments. Fine Content Indicates Clay and Silt

	Lake Sanaru	Shimizu Port	Osaka Bay	Seto Island
Specific Gravity	2.50–2.67	2.56–2.61	2.54–2.69	2.67
Water Content (%)	125–407	66–173	154–224	26.6
Void ratio	3.33–10.2	1.72–4.42	391–6.03	0.71
Porosity (%)	76–1	63.2–81.5	79–85.8	41.5
Fine Content	85–99	48–70	98–99	(sandy)

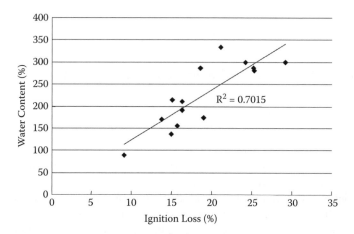

FIGURE 2.10　Relationship between water content and ignition loss for surface sediments in a port area.

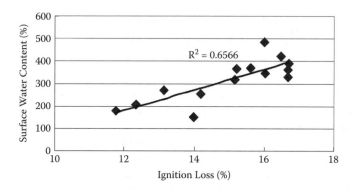

FIGURE 2.11　Relationship between water content and ignition loss for a core sample.

2.4　BENTHOS

Benthos are the organisms which live on, in, or near the bottom of lakes, rivers, streams, and seas. Many types of benthos ingest the organic matter of sediments. This contributes to the purification of water and sediments. On the other hand, benthos are influenced by the contamination of sediments. Therefore, benthos can be an index of sediment quality (Hale and Heltshe, 2008). In this sense, when the quality of surface water is concerned, benthos and their diversity are regarded as components of sediment quality (McPherson et al., 2008).

According to size, benthos are classified as macrobenthos, meiobenthos, nanobenthos, picobenthos, etc. Macrobenthos are organisms larger than 1 mm. They include polychaete worms, pelecypods, echinoderms, sponges, ascidians, and crustaceans. In freshwater ecosystems, the benthic macroorganisms provide a good visual indicator of water quality.

Meiobenthos (meiofauna) are a group of organisms larger than microbenthos (microfauna) but smaller than macrobenthos (macrofauna). The name is, therefore, not a taxonomic grouping. In practice, this category includes organisms that can pass through a 1-mm mesh but are retained by a 45-μm mesh.

Microbenthos include nanobenthos and picobenthos and are microscopic benthos that are less than 32 μm in size. They include bacteria, diatoms, ciliates, amoebas, and flagellates.

It is important to note that benthos play two important roles in environmental problems. Benthos may be part of the food chain, which is different from that in the surface water. A high concentration of contaminants can accumulate in the food chain. Another is sediment disturbance (bioturbation), which can cause environmentally beneficial or adverse effects. The disturbance can supply oxygen into the deeper sediments. On the other hand, the contaminants in surface sediments can be transported into the subsoil by the disturbance.

2.5   USES OF SEDIMENTS AND WATER

Sediments and water are components of the earth, which can be called a planet of water. This is not only because water covers the surface of the earth, but also water is the most fundamental substance required for living things including humans. The circulation of water acts just like the blood for the earth. That is why the water quality should be preserved. Contamination of water may cause a disruption of the life cycle due to biological accumulation or biological concentration through food-chain processes. All aquatic life exists because there has been water there. Environmental sustainability cannot be obtained without water.

Aquaculture has been one of the newer types of fisheries. For these activities, water quality is one of the most important factors. Because of eutrophication, red and blue tides often kill fish in the enclosure pens. In addition, aquaculture is perceived by governments and international agencies as an economic alternative for poor communities in developing countries. Nevertheless, aquaculture must address environmental issues as well as economic and social aspects to achieve sustainable development (Rodríguez-Gallego et al., 2008). As long as the water quality is kept safe, it can be used for the aquaculture of shrimp and fish on land (Avnimelech, 2006; McLachlan et al., 2001).

Sediments supply a place for microbial activities, soils for water plants, and the habitats of benthos and aquatic life. On the other hand, the food-chain process originates in the sediments (Bright et al., 1995). The organic materials in sediments are food for benthos and crustacea. Sandy beaches are also sediments which provide a habitat for a variety of life and a place of recreation for people. The biodiversity in sediments is key for a sustainable ecosystem in a water area.

Freshwater sediments have been used as soils for agriculture. The organic-rich sediments in rivers and lakes are suitable for farms and paddy fields. However, it is noted that sediments are sometimes contaminated. The leaching tests for sediments are an insufficient method for evaluation of the contamination of sediments because some contaminants, such as heavy metals, are strongly adsorbed onto soil particles. The heavy metals adsorbed onto particles cannot be released under normal conditions.

The contaminated sediments still have an adsorption capacity for heavy metals (Fukue et al., 2001). Although adsorbed metals may not be easily released, some plants can release acid from their roots, thus releasing the metal ions from the particles. Consequently, the plants can absorb the released contaminants. This occurred in Toyama Prefecture, Japan, in 1950. The event, which is called itai-itai disease, was the first documented case of mass cadmium poisoning in the world (http://en.wikipedia.org/wiki/Itai-itai_disease). The cadmium was released in to the rivers by mining companies in the mountainous region. An earlier investigation was made on the water, but no cadmium was detected. It was unfortunate that no cadmium was detected from the leaching tests of sediments. It is noted that, in this case, the quality of sediments can be evaluated by means of metal content tests, but not a leaching test.

Although marine sediments cannot be used as farm soils, because of the salt content, they have been used as reclamation materials in coastal regions. Dredging materials have often been used as construction materials. In 1970s, dredging with water pumps was used, and pipe lines were used to transport the dredged materials. The length of the pipe lines often exceeded several kilometers.

To maintain the depth of navigation routes for ships, dredging has often been performed in ports. At present, the dredged materials are usually contaminated and regarded as waste materials. Therefore, the dredged materials usually have to be disposed of in designated areas.

2.6   MANAGEMENT OF SEDIMENTS

Sediments and benthos are often considered to have a substantial influence on water quality. This is because water quality cannot be considered without the quality of sediments. This is why capping with sand on sediments has been used for the control of water quality.

To preserve the quality of surface water, the sediment quality should also be preserved. Contaminated sediments will release contaminants or nutrients through the diffusion or dissolution of organic matter and carbonates, or other processes. These natural processes are difficult to control. This is a main reason why a variety of proposals for the control of water quality have failed. Much effort has been made to reduce the chemical oxygen demand (COD) of water. To achieve this, capping, dredging, and oxidation techniques have been applied in many cases.

Capping criteria have been discussed by Mohan et al. (2000). They indicated that in situ, subaqueous capping is an attractive, nonintrusive, and cost-effective method of remediating contaminated sediments. The successful design of a subsurface cap requires the proper application of hydraulic (armor and filter equations), physical (diffusive and advective/dispersive transport equations), and geotechnical (settlement and stability equations) engineering principles. Theoretical considerations for cap and armor design were presented in the paper and illustrated using an application of a cap design for a confidential contaminated harbor site.

2.7 CONCLUDING REMARKS

The quality of surface water depends on sediment quality including biodiversity in sediments. This is because sediment quality strongly influences the quality of surface water through interfacial phenomena. In particular, the degradation of organic matter will impact water quality. Therefore, the content of organic matter, which is also related to eutrophication, will control the sediment environment.

The quality of sediments can be described by two aspects; one is "eutrophication" (i.e., mainly the content of nitrogen and phosphorus), and another is "contamination due to toxic substances." Therefore, the management and control of sediments concerns the reduction of nutrients and contaminants in an economical way from the water system, including sediments. To achieve this, general knowledge in sedimentology, marine geology, geochemistry, surface chemistry, geotechnical engineering, microbiology, and aquatic biology is required.

REFERENCES

Avnimelech, Y. 2006. Bio-filters: The need for a new comprehensive approach. *Aquacult. Eng.* 34: 172–178.

Bright, D.A., Dushenko, W.T., Stephen, L., Grundy, S.L., and Reimer, K.J. 1995. Effects of local and distant contaminant sources: polychlorinated biphenyls and other organochlorines in bottom-dwelling animals from an Arctic estuary. *Sci. Tot. Environ.* 160/161: 265–283.

Cappuyns, V., Swennen, R., and Devivier, A. 2006. Dredged river sediments: potential chemical time bombs?: A case study. *Water Air Soil Poll.* 171: 49–66.

Fabris, G.J., Monahan, C.A., and Batley, G.E. 1999. Heavy metals in water and sediment of Port Philip Bay, Australia. *Aust. J. Mar. Freshwater Res.* 50: 503–513.

Fukue, M. and Mulligan, C.N. 2009. Development of a theoretical approach for prediction of soil compression behaviour. *Soils Found.* 49(1): 99–114.

Fukue, M., Mulligan, C.N., Sato, Y., and Fujikawa, T. 2007. Effect of organic suspended solids and their sedimentation on the surrounding sea area. *Environ. Poll.* 149(1): 70–78.

Fukue, M., Yanai, M., Sato, Y., Fujikawa, T., Furukawa, Y., and Tani, S. 2006. Background values for evaluation of heavy metal contamination in sediments. *J. Haz. Mat.* 136: 111–119.

Fukue, M., Yanai, M., Takami, Y., Kuboshima, S., and Yamasaki, S. 2001. Containment, sorption, and desorption of heavy metals for dredged sediments. In K. Adachi and M. Fukue (Eds.), *Clay Science for Engineering*. Balkema, Rotterdam, pp. 389–392.

Hale, S.S. and Heltshe, J.F. 2008. Signals from the benthos: Development and evaluation of a benthic index for the nearshore Gulf of Maine. *Ecol. Indicat.* 8: 338–350.

Krumbein, W.C. and Sloss, L.L. 1963. *Stratigraphy and Sedimentation.* 2nd ed. W. H. Freeman and Company, San Francisco, 660 pp.

McLachlan, M.C., Haynes, D., and Müller, J.F. 2001. PCDDs in the water/sediment-seagrass-dugong (*Dugong dugon*) food chain on the Great Barrier Reef (Australia). *Environ. Poll.* 113: 129–134.

McPherson, C., Chapman, P.M., deBruyn, A.M.H., and Cooper, L. 2008. The importance of benthos in weight of evidence sediment assessments—A case study. *Sci. Total Environ.* 394: 252–264.

Mohan, R.K. 2000. Modeling the physical and chemical stability of underwater caps in ports and harbors. In Herbich, J.B. (Ed.), *Handbook of Coastal Engineering*, Chapter 14. McGraw-Hill Professional, New York, pp. 14.1–14.27.

Rodríguez-Gallego, E., Meerhoff, L., Poersch, L., Aubriot, L., Fagetti, C., Vitancurt, J., and Conde, D. 2008. Establishing limits to aquaculture in a protected coastal lagoon: Impact of *Farfantepenaeus paulensis pens* on water quality, sediment and benthic biota. *Aquaculture* 277: 30–38.

Wen, B, Aydin, A., and Duzgoren-Aydin N.S. 2002. A comparative study of particle size analysis by sieve-hydrometer and laser diffraction methods. *Geotechnical Testing Journal, ASTM International* 25(4): 434–442.

Wetzel, A. 1989. Influence of heat flow on ooze/chalk concentration; Quantification for consolidation parameters in DSDP sites 504 and 505, sediments. *J. Sediment Petrol.* 59: 539–547.

Yong, R.N. and Mulligan, C.N. 2004. *Natural Attenuation of Contaminants in Soils*. CRC Press, Boca Raton, FL.

Yong, R.N., Mulligan, C.N., and Fukue, M. 2006. *Geoenvironmental Sustainability*. CRC Press, Boca Raton, FL.

3 Contaminant–Sediment Interactions

3.1 INTRODUCTION

As discussed in the previous chapter, sediments consist of solids, pore water, and gases. The interactions of the pore water with the contaminants and the solids are complex and are to be discussed in this chapter. It is important to understand the physical and physicochemical interactions of the contaminants with the sediment solids to understand the capacity of retention of the sediments and potential parameters for contaminant release. Sediment composition, properties, and characteristics will influence the interactions at the sediment–pore water interface. The reactions between a pollutant and sediment will determine its transport through the sediments, and also its fate.

3.2 FACTORS INFLUENCING CONTAMINANT–SEDIMENT INTERACTIONS

3.2.1 SPECIFIC SURFACE AREA (SSA)

The sediment fractions that have more particles with significant reactive surfaces are the clay minerals, oxides and hydrous oxides, soil organics, and carbonates. Table 3.1 gives the surface charge characteristics, specific surface area (SSA), and cation exchange capacity (CEC) for some clay minerals. SSA is the total surface area of all the solids or particles per unit volume or unit weight. Determination of the amount of gas or liquid (adsorbate) that forms a monolayer coating on the surface of the particles is often used to evaluate SSA. The choice of the adsorbate and the availability of sediment particles in a totally dispersed state are important. The surface charge density is the total number of electrostatic charges on the particles' surfaces divided by the total surface area of the particles.

The transmission property of sediment refers to those properties which participate in the transport of pore water and other fluids. This is most important for coarse-grain sediments (Burdige, 2006). These are essentially described by the permeability of the sediment, related mainly to the aqueous phase, that is driven by pressure differences due to wave action or bottom currents. For fine-grained sediments with low permeability, diffusion and bioturbation are more important pore water transport processes. The factors that affect hydraulic conductivity are divided into two groups: (a) those that pertain to the fluid phase and (b) those that concern the solid particles and structure. A unit sediment mass is made up of an almost infinite number and

TABLE 3.1

Charge Characteristics, SSA, and CEC for Some Clay Minerals

Clay Mineral	Cation Exchange Capacity (CEC), meq/100 g	Surface Area, m^2/g	Range of Charge meq/100 g
Kaolinite	5–15	10–15	5–15
Clay micas and chlorite	10–40	70–90	20–40
Illite	20–30	80–120	20–40
Montmorillonite	80–100	800	80–100
Vermiculite	100–150	700	100–150

Source: Adapted from Yong, 2001.

arrangement of particles and peds (fabric units), as shown in Figure 3.1. Accordingly, sediments with similar compositions can have different densities and correspondingly different hydraulic conductivities and could also possess different hydraulic conductivities due to various macro- and microstructures.

Permeability to water (aqueous phase) is measured as hydraulic conductivity and is commonly expressed in terms of a Darcy permeability coefficient, k, which is generally obtained via application of the Darcy model as a means of analysis of hydraulic conductivity data. From laboratory permeability measurements, the Darcy

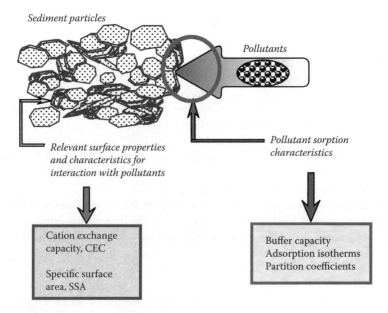

FIGURE 3.1   Some relevant sediment properties for interaction with contaminants. The sediment structure consists of dispersed particles which are typical of sediment suspensions used for determination of adsorption isotherms and compact microstructures.

coefficient k is obtained from the relationship: $v = ki = k(\Delta h/\Delta L)$. The hydraulic gradient i is the ratio of the hydraulic head Δh and ΔL, the spatial distance.

To incorporate the influence of the permeant and sediment structure properties such as permeant viscosity, sediment-voids' features, tortuosity, and shape of pore space cross section, a different type of permeability coefficient will be needed. To obtain this new coefficient, the Poiseuille relationship for flow through fine-bore tubes shown in Equation (3.1) is used in an adapted form for determination of the link between structure and permeability.

$$v^* = \frac{r^2\gamma}{8\eta}\frac{\Delta\psi}{\Delta l} \tag{3.1}$$

where:
v^* = mean effective flow velocity through a narrow tube of radius r
γ and η = density and viscosity of fluid or permeant
$\Delta\psi$ = potential difference between the ends of a tube of length Δl

To account for the influence of the properties of pore channels defined by the structure of a sediment and the fact that the wetted sediment particles' surface area is controlled by the structure of the sediment, Yong and Mulligan (2002) used a modification of the combined form of the Poiseuille and Kozeny-Carman model—as shown in Equation (3.2).

$$v = k^* i = \frac{C_s n^3 \gamma}{\eta T^2 S_w^2}\frac{\Delta\psi}{\Delta l} \tag{3.2}$$

where:
k^* = permeability coefficient which considers permeant and sediment structure properties.

$$\frac{C_s n^3 \gamma}{\eta T^2 S_w^2}$$

C_s = shape factor and has values ranging from 0.33 for a strip cross-sectional face to 0.56 for a square face. This factor accounts for the fact that the cross-sectional face of any of the pore spaces in the sediment mass is highly irregular and allows one to choose a typical value for a representative pore cross section area. Yong and Warkentin (1975) have suggested that a value of 0.4 for C_s may be used as a standard value—with a possible error of less than 25% in the calculations for an applicable value of k^*.
i = hydraulic gradient = ratio of the potential difference $\Delta\psi$ between the entry and exit points of the permeant and the direct path length Δl of the sediment mass being tested.

T = tortuosity = ratio of effective flow path Δl_e to thickness of test sample Δl and which is quite often taken to be $\approx \sqrt{2}$.

γ and η = density and viscosity of the permeating fluid respectively.

n = porosity of the unit sediment mass.

S_w = wetted surface area per unit volume of sediment particles.

Equation (3.2) contrasts with the standard Kozeny-Carman (K-C) relationship in that the wetted surface area consideration in the K-C model assumes that $S_w = S(1-n)$ and that the radius r of the Poiseuille tube is:

$$r = \frac{n}{S(1-n)}$$

This gives the relationship for k from the K-C model as:

$$k = \frac{C_s \gamma n^3}{\eta T^2 S^2 (1-n)^2} \qquad (3.3)$$

where S = specific surface area of sediment. With this measure of surface area, all particle surfaces are assumed to be in contact with the permeating fluid.

The adaptation introduced by Yong and Mulligan (2002) considers the surface area of the particles in terms of only the wetted surfaces (S_w). This allows consideration of the compositional differences and sediment structure differences which can impact severely on the distribution of pore sizes and availability of sediment particle surfaces for direct interaction with the permeant. It can be seen from Equation 3.2 that C_s, T, and S are sediment property parameters which are dependent on sediment composition and structure. These can be expressed as a parameter $\beta = C_s/(TS)^2$. Along the same lines, the density and viscosity of the permeating fluid, γ and η, respectively, are properties of the permeant and can be described by a parameter $\mu = \gamma/\eta$. Using these parameters, the relationship for k^* can be expressed as follows: $k^* = \mu\beta n^3$. Assuming that the physical properties of a permeant are close to water at about 20°C, μ can be computed directly. Further assuming a tortuosity T value of $\sqrt{2}$, and $C_s = 0.4$, the graphical relationships shown in Figure 3.2 can be obtained. This graph shows the relationship between the sediment permeability expressed as a coefficient k^* and the amount of surface area wetted in fluid flow through the sediment—all of which are determined in relation to the porosity of the sediment. The wetted surface area is the surface area through which fluid flow occurs. The wetted surface area S_w is a small fraction of the specific surface area of the various sediments. If the surface area of the solids is comprised of a unit volume of the sediment being permeated as SSA_v, the ratio S_w/SSA_v can be defined as the wetted surface ratio (WSR). The WSR provides an indication of the microstructure of the sediment, the extent of particle surfaces available for interaction with the fluid used for the permeability test, and thus the surface area wetted during hydraulic flow (Yong and Mulligan, 2002). It will be shown in Chapter 7 how this property can affect heavy metal removal.

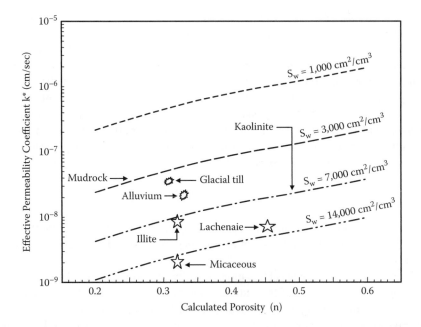

FIGURE 3.2 Relationship of the permeability coefficient k* with calculated porosity. The effect of wetted surface area S_w is also shown (adapted from Yong and Mulligan, 2002).

3.2.2 CATION EXCHANGE CAPACITY (CEC)

Interactions occurring between the pollutants in the pore water (dissolved solutes) and reactive particle surfaces are responsible for the transfer of these solutes from the pore water to the sediment surfaces (partitioning). Molecular interactions governing sorption of pollutants are electrostatic in nature. They are coulombic interactions between nuclei and electrons. Of particular importance are the interatomic bonds such as the ionic, covalent, hydrogen, and van der Waals. Ionic forces are coulombic forces. These are forces between positively and negatively charged atoms, and the bonds formed are called ionic or electrovalent bonds. The simplest example of ionic bonding is between a sodium atom and a chlorine atom—resulting in the formation of NaCl. The strength of the attractive forces and, hence, the strength of the ionic bonds decrease as the square of the distance separating the atoms.

Cation exchange in sediments refers to the exchange of positively charged ions associated with clay particle surfaces. The process is stoichiometric, and electroneutrality at the clay particle surfaces must be satisfied. Cations are attracted to the reactive sediment particle surfaces in accordance with the relationship

$$\frac{M_S}{N_S} = \frac{M_O}{N_O} = 1$$

where M and N represent the cationic species, and the subscripts s and o represent the surface and the bulk solution, respectively.

Exchangeable cations are cations that can be readily replaced by other cations of equal valence, or by two of one-half the valence of the original one to maintain the balance of charge. Thus, in freshwater sediments, if a clay containing sodium as an exchangeable cation is washed with a solution of calcium chloride, each calcium ion will replace two sodium ions, and the sodium ions can be washed out in the solution. In marine sediments, on the other hand, for seawater where there is a high concentration of sodium ions, the opposite, uptake of Na^+ and release of Ca^{2+} can occur (Sayles and Manglesdorf, 1977). This is particularly important when fine particles are transported from freshwater rivers to the ocean or for ocean dumping (Burdige, 2006). The reaction can be represented as:

$$NA_o^{M+} + MB_s^{N+} \leftrightarrow MB_o^{N+} + NA_s^{M+} \tag{3.4}$$

The subscript o denotes the aqueous phase ions, whereas the s represent the solid phase ions that are adsorbed or at cation exchange sites. M and N refer to the valence of the cations.

The quantity of exchangeable cations held by the sediment is called the cation exchange capacity (CEC) of the sediment and is expressed as milliequivalents per 100 g of sediment (meq/100 g sediment). The CEC is a measure of the amount of negative sites associated with the sediment fractions. The predominant exchangeable cations in sediments are calcium and magnesium, with smaller amounts of potassium and sodium. The valence of cations plays a significant role in the exchange process. Higher valence cations will show greater replacing power. The higher the charge, the higher is its attraction to exchange sites. An example of some typical cations and the replacing power is given as follows:

$$Th^{4+} > Fe^{3+} > Al^{3+} > Cu^{2+} > Ba^{2+} > Ca^{2+} = Mg^{2+} > Cs^+ > K^+ = NH_4^+ > Li^+ > Na^+$$

Exchange-equilibrium equations can be used to determine the proportion of each exchangeable cation to the total cation exchange capacity (CEC) as the outside ion concentration varies. The simplest is the Gapon relationship:

$$\frac{M_e^{m+}}{N_e^{n+}} = K \frac{\left[M_o^{m+} \right]^{\frac{1}{m}}}{\left[N_o^{n+} \right]^{\frac{1}{n}}} \tag{3.5}$$

where m and n refer to the valence of the cations, and the subscripts e and o refer to the exchangeable and bulk solution ions, respectively. The constant K is dependent on the effects of specific cation adsorption and the nature of the clay surface. K decreases in value as the surface density of charges increases. The adsorption of ions due to the mechanism of electrostatic bonding is called physical adsorption or nonspecific adsorption.

The surface properties of sediments are important, because it is these properties, together with those surface properties of pollutants themselves and the geometry and continuity of the pore spaces that will control the transport processes of the pollutants. The sediment fractions that possess significant reactive surfaces include layer silicates (clay minerals), organic matter, hydrous oxides, carbonates, and sulfates. The surface hydroxyls (OH^- group) are the most common surface functional group in inorganic sediment particles such as clay minerals with disrupted layers (e.g., broken crystallites), hydrous oxides, and amorphous silicate minerals. The common functional groups for organic matter include the hydroxyl, carboxyl, and phenolic groups and amines.

3.3   SORPTION OF POLLUTANTS AND PARTITION COEFFICIENTS

The processes of transfer of metal cations from the sediment pore water can be grouped as follows. Sorption includes physical adsorption (physisorption), occurring principally as a result of ion exchange reactions and van der Waals forces, and chemical adsorption (chemisorption), which involves short-range chemical valence bonds as previously discussed. The term sorption is used to indicate the process in which the solutes (ions, molecules, and compounds) are partitioned between the liquid phase and the particle interface. As it is difficult to fully distinguish between the mechanisms of physical adsorption, chemical adsorption, and precipitation, the term sorption is used.

Physical adsorption occurs when the pollutants or contaminants in the solution (aqueous phase, pore water) are attracted to the surfaces of the sediment particles because of the unsatisfied charges. In the case of the heavy metals (metal cations) for example, they are attracted to the negative charges exhibited by the surfaces of the particle solids. This sorption is a function of pH. For example, $Fe(OH)_3$, a major soil component, has a variable charge with pH. The pH of the zero charge point for this component, where the positive and negative charges are equal, is 8.5. Below that pH, cationic species would be unlikely to sorb onto the cationic surface.

In the case of heavy metals, precipitation of the heavy metals will also remove the heavy metals from solution. Precipitation mechanisms for organic chemical pollutants usually do not occur, so it is generally assumed that the total "partitioned" organic chemicals are sorbed or attached to the solids. The partitioning or distribution of the organic chemical pollutants is described by a coefficient identified as k_d. As defined previously, this coefficient refers to the ratio of the concentration of pollutants "held" by the sediment fractions to the concentration of pollutants "remaining" in the pore water (aqueous phase), i.e., $C_s = k_d C_w$, where C_s refers to the concentration of the organic pollutants sorbed by the sediment fractions, and C_w refers to the concentration remaining in the aqueous phase (pore water), respectively. Therefore, sediment chemistry and surface characteristics, redox potential, pH, and speciation of the contaminant will all influence sorption.

The equilibrium partitioning of pollutants refers to the steady state of transfer of organic pollutants from the pore water to the sediment solids or the reverse, where there is desorption of the pollutant from the sediment particles by processes that include all of those previously described. Organic matter in the pore water and adsorbed onto the sediments can play a significant role. Determination of partitioning of inorganic contaminants and pollutants is generally conducted using batch equilibrium tests. Results obtained from the tests are called adsorption isotherms.

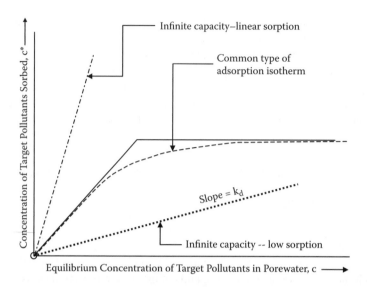

FIGURE 3.3 Partitioning of pollutants between the pore water and pollutants sorbed by sediment particles.

The three common types of adsorption isotherms (Freundlich, Langmuir, and constant) are shown in Figure 3.3. The parameter k_d in the equations shown with the various curves denotes the slope of the curves.

Organic matter exists as dissolved and suspended forms and on the bottom sediments. The functional groups of the organic matter interact with heavy metals. The affinities of these groups for heavy metals in decreasing order are:

enolates > amines > azo compounds > ring N > carboxylates > ethers > carbonyls

On the other hand, organic matter may lead to the extraction of the heavy metals via mineral dissolution and solubilization of metal sulfides and carbonates. The factors of pH, alkalinity, redox potential, and amount of organic matter can all influence the sorption of heavy metals (USEPA, 2005). High levels of Ca, Na, Mg, and K may also decrease heavy metal sorption.

Volatilization may be an important attenuation mechanism for volatile organic contaminants. Freshly spilled petroleum products such as gasoline can exhibit high rates of volatilization that can occur from the free phase or dissolved phase. Henry's constant law describes volatilization from the dissolved phase. The rate of volatilization slows as the age of the spill increases. As a general guideline, a dimensionless Henry's constant greater than 0.05 means that volatilization or off-gassing is likely, while if it is less than 0.05, volatilization would be negligible. In sediments, this mechanism is not a dominant one due to the depth of the sediments in the water column.

However, some components such as mercury can be subjected to volatilization from the surface water (Morel et al., 1998). In lakes, sedimentation and volatilization are major mechanisms of mercury loss, while a number of biochemical and chemical reactions can occur in oceans. Precipitation–volatilization and oxidation–reduction

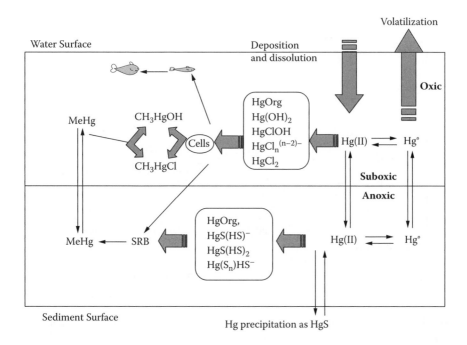

FIGURE 3.4 Mechanisms of Hg conversion in surface water. SRB = sulfate-reducing bacteria. (Adapted from Morel et al., 1998).

reactions function in the mercury cycle. Due to atmospheric inputs of mercury, levels in the sediment have accumulated over the past 150 years (Mason et al., 1994). The forms of Hg can be seen in Figure 3.4. There is elemental mercury, which is volatile but relatively stable, and various mercury species. Near the air–water interface, Hg dominates, whereas total Hg and methyl mercury dominate near the sediments. Total mercury includes particulate and soluble species.

Chemical mass transfer is responsible for partitioning of contaminants in the fate and transport of contaminants. Reduction–oxidation reactions can also play an important role in the fate of the contaminants. Assessment of the retention or retardation processes is required to understand partitioning and the attenuation of contaminants within the sediment. If potential pollution hazards and threats to public health and the environment are to be minimized or avoided, we must ensure that the processes for contaminant attenuation are irreversible and the levels of contaminants are below allowable limits or levels.

For example, for arsenic, two models exist in respect to possible mechanisms for release of arsenic from the arsenic-bearing materials, as shown in Figure 3.5: (a) reduction mechanisms and (b) oxidation processes. In the former process, it is reasoned that reductive dissolution of arseniferrous iron oxyhydroxides releases the arsenic responsible for pollution of the groundwater. The other model for arsenic release from the alluvium relies on oxidation of the arsenopyrites as the principal mechanism. This occurs when oxygen invades the groundwater because of the lowering of the groundwater from the abstracting tubewells. The sorption of arsenic (III)

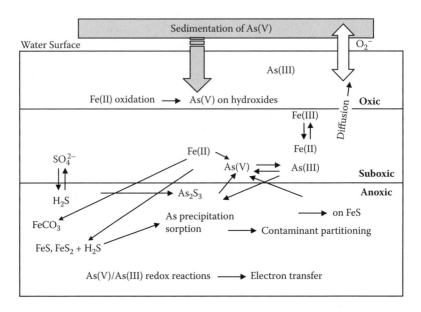

FIGURE 3.5 Mechanisms of As conversion in surface water (adapted from Bostick et al., 2004).

by anoxic estuarine sediments has been studied by Bostick et al. (2004). Although sorption was apparent at all pH values, it was more significant at pH 7. Sorption conformed to Langmuir isotherms. Iron sulfide fractions were responsible for most of the sorption. In addition, over time, the FeAsS-like precipitates reacted to form As_2S_3 and, when combined with the drop in redox potential, stabilized the arsenic. The sorbed species of arsenic were determined by extended X-ray absorption fine structure (EXAFS) spectroscopy.

The organic matter of sediments may change in structure, thus binding metals and other chemicals more tightly (Pignatello et al., 1993). Contaminants such as heavy metals may diffuse into the sediment structure and thus may be tightly bound as well (Steinberg et al., 1987). Petroleum compounds over time lose the more soluble and volatile components (Wilcock et al., 1996) and are thus less bioavailable and less biodegradable (Sandoli et al., 1996).

3.3.1 PARTITIONING OF INORGANIC POLLUTANTS

Partitioning of inorganic and organic chemical pollutants is often represented by the partition coefficient k_p. In brief, partition coefficients describe the relationship between the amount of pollutants transferred onto sediment particles and the equilibrium concentration of the same pollutants remaining in the pore water (Figure 3.6). The popular relationships such as Langmuir and Freundlich are shown. Partitioning is the result of mass transfer of pollutants from the pore water. There are at least two broad issues regarding the determination and use of the distribution coefficient k_d, namely: (a) types of tests used to provide information for determination of k_d and (b)

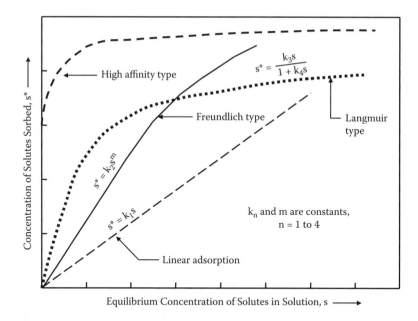

FIGURE 3.6 Different types of adsorption isotherms obtained from batch equilibrium tests.

range of applicability of k_d in transport and fate predictions. Laboratory tests used to provide information on the mass transfer of pollutants from the pore water onto sediment solids are the most expedient means to provide one with information on the partitioning of pollutants. By and large, these tests provide only the end result of the mass transfer, and not direct information on the basic mechanisms responsible for partitioning.

The distribution coefficient k_d is determined from information gained using batch equilibrium tests on sediment solutions. Ratios of 10 or 20 parts of solution to one part sediment are generally used, and the candidate or target pollutant is part of the aqueous phase of the sediment solution. In many laboratory test procedures, the candidate sediment is used for the solid in solution, and the candidate or target pollutant is generally a laboratory-prepared pollutant, such as $PbNO_3$ for assessment of sorption of Pb as a pollutant heavy metal. Since the sediment particles are in a highly dispersed state in the slurry, the surfaces of all the particles are available for interaction with the target pollutant in the aqueous phase of the solution. By using multiple batches of sediment solution where the concentration of the target pollutant is varied, and by determining the concentration of pollutants sorbed onto the sediment solids and remaining in the aqueous phase, the characteristic adsorption isotherm curve is obtained, and the slope is defined as k_d as shown in Figure 3.6.

Distribution coefficients k_d obtained from adsorption isotherms using the batch equilibrium with sediment solutions and prepared target pollutants are very useful in that they define the upper limit of partitioning of the target pollutant.

The aging or changes over time of the sediments and/or contaminants is referred to as weathering. The dissolution of metal sulfides can release metals such as zinc and lead into the environment. Temperature, surface area of the solids, pH, particle size distribution, oxygen levels, the water flow rate, and ionic strength can all influence weathering rates. Bentley et al. (2006) showed that laboratory column studies could be used to relate the lab studies to field dissolution rates. Scale factors are needed to predict and relate bulk physicochemical lab and field sites. These types of information are important for evaluating metal release from dredged sediments and movement of sediments in the water column.

For assessment of partitioning using sediments in their natural state, it is necessary to conduct column-leaching or cell-diffusion tests. In these kinds of tests, the natural sediment is used in the test cell or column, and either laboratory-prepared candidate pollutants or natural leachates are used. The partition coefficient deduced from the test results is not the distribution coefficient identified with the adsorption isotherms obtained from batch equilibrium tests. Instead, the partition coefficients obtained from column-leaching or cell-diffusion tests need to be properly differentiated from the traditional k_d. Yong (2001) suggested that these partition coefficients be called sorption coefficients to reflect the sorption performance of the soils in their natural state in the column or cell. The disadvantages in conducting column-leaching and cell-diffusion tests are (a) the greater amount of effort required to conduct the tests, (b) the much greater length of time taken to obtain an entire suite of results, and (c) inability to obtain exact replicate soil structures in the companion columns or cells. The results indicate that the characteristic curves obtained from column-leaching tests, for example, are much lower than corresponding adsorption isotherms. Figure 3.7 gives an example of an experimental setup to perform these tests.

FIGURE 3.7 Leaching column setup.

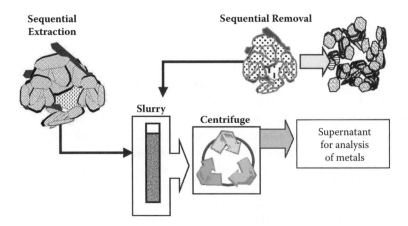

FIGURE 3.8 Methodology for SSE tests.

3.3.2 SELECTIVE SEQUENTIAL EXTRACTION

Measurement of the mobility and availability of metals is required to predict and interpret their behavior. As total metal concentrations do not give a good indication of metal toxicity, other methods are needed. Trace metals can be found in numerous sediment and soil components in different ways (Krishnamurthy et al., 1995). Metals in river sediments can be bound to different compartments: adsorbed onto clay surfaces or iron and manganese oxyhydroxides; present in the lattice of secondary minerals like carbonates, sulfates, or oxides; attached to amorphous materials such as iron and manganese oxyhydroxides; and complexed with organic matter or in the lattice of primary minerals such as silicates (Gismera et al., 2004; Schramel et al., 2000; Tessier et al., 1979). To determine the fractionation of metals in soils, various methods are used. One method is to use specific extractants called selective sequential extraction. By sequentially extracting with solutions of increasing strength, a more precise evaluation of the different fractions can be obtained (Tessier et al., 1982). A soil or sediment sample is shaken over time with a weak extractant and centrifuged, and the supernatant is removed by decantation (Figure 3.8). The pellet is washed in water, and the supernatant is removed and combined with the previous supernatant. A sequence of reagents is used following the same procedure until, finally, mineral acid is used to extract the residual fraction. Heavy metal concentrations are then determined in the various extracts by atomic absorption, inductively coupled plasma (ICP), or other means. Numerous techniques and reagents have been developed and have been applied to soils (Shuman, 1985), sediments (Tessier et al., 1982), sludge-treated soils (Petrozelli et al., 1983), and sludges.

Although none of the extractions is completely specific, the extractants are chosen to minimize solubilization of other fractions and provide a distribution of the partitioning of the heavy metals. The extracting agents increase in strength throughout the sequence to destroy the bonds of the heavy metals to the various sediment components of increasing strength (Yong, 2001). As an example, Koeckritz et al. (2001) proposed an equivalent step to simplify the sequential extraction procedure designed by Zeien and Brummer (1989). They reduced four initial steps in the procedure to one with no significant change in the results.

Ammonium acetate, barium chloride, or magnesium chloride at pH 7.0 is generally used to extract the exchangeable fraction by displacement of the ions in the sediment matrix bound by electrostatic attraction (Lake, 1987). Calcium chloride, potassium nitrate, and sodium nitrate can also be used (Yong, 2001). Hydroxylamine hydrochloride with acetic acid at pH 2.0 reduces the ferrous and manganese hydroxides (reducible phase) to soluble forms (Tessier et al., 1979). The carbonate phase (calcite and dolomite) is extracted at pH 5.0 with sodium acetate acidified with acetic acid by solubilization of the carbonates, releasing the carbonate-entrapped metals (Yong and Mulligan, 2004). Hot hydrogen peroxide in nitric acid is used to oxidize the organic matter, thus releasing the metals that are complexed, adsorbed, and chelated. The silicates should not be affected by this treatment (Yong, 2001). In the final step, strong acids at high temperatures dissolve the silicates and other materials. This residual fraction is usually used to complete the mass balances for the metals. Yong et al. (1999) reported that, through selective sequential extraction techniques (SSE) described in Table 3.2, the strength of retention mechanisms of heavy metals by the phases of solids decreased in the following order:

carbonates > amorphous > organics > exchangeable

Ho and Evans (2000) investigated the mobility of heavy metals through SSE methods with assessment of readsorption effects. The study showed that Cd was highly mobile, Cu and Pb were associated primarily with oxidizable organic matter, and Zn was found in all fractions. Chartier et al. (2001) indicated that 18% to 42% of Pb, Zn, and Cd exist in the carbonate-bound fraction, while 39% to 60% of these metals were associated with the iron and manganese oxide bound fraction. The study also showed that 65% to 72% of total copper present in the sediments was found in organic matter and sulfide bound fractions; 50% to 80% of Ni and Cr in sediment exist in the residual fraction.

TABLE 3.2
Reagents Used for Sequential Extraction Procedure

	Chemical Reagents	Fraction
1	Water or surfactant	Soluble
2	$MgCl_2$ (pH 7)	Exchangeable
3	NaOAc (pH 5 with acetic acid)	Carbonates
4	$NH_2OH \cdot HCl$ in 25% (v/v) acetic acid (pH 2.5)	Oxides and hydroxides
5	HNO_3 and 30% H_2O_2 (pH 2), 30% H_2O_2 (pH 2) NH_4OAc in 20% (v/v) HNO_3	Organic matter
6	Aqua regia (HCl, HNO_3, and water)	Residual

Note: Ac—denotes acetate.

TABLE 3.3

Sequential Extraction of Metals for Two Sediment Samples

Metal	Fraction (% Of Total)				
	Exchangeable	Carbonate	Oxide	Organic	Residual
Lachine Canal					
Copper	1	1	4	86	12
Nickel	0	9	23	29	39
Zinc	4	18	46	22	10
Lake Sanaru					
Copper	<1	<1	<1	73	27
Lead	20	9	23	20	28
Zinc	9	10	28	24	28

Samples from Lachine Canal, Montreal, Canada, and Lake Sanaru, Japan, were evaluated by the SSE test. The results are shown in Table 3.3. For both samples, the major fractions are the oxide fraction, which was dominant for zinc, and the organic fraction, which constituted more than 70% of the total copper. The residual fraction (which is the most stable fraction) seemed to hold about 30% of zinc and 30% of copper for the Lake Sanaru sample, whereas the levels were lower in the Lachine Canal sediment. Only lead in the Lake Sanaru sample showed a significant ion exchangeable fraction, which is the most likely to desorb from the sediment.

Another important aspect is that SSE can be used for the evaluation of a proposed removal technique for specific heavy metals. Peters (1999) reported that SSE tests show that Cu, Pb, Zn, and Cr existed mostly in the amenable fractions (i.e., exchangeable + carbonate + reducible oxide). Mulligan and Dahr Azma (2003) showed that sequential extraction can be employed for the evaluation of the most appropriate sediment remediation technology and for monitoring remediation procedures. This will be discussed further in Chapter 6. A rhamnolipid biosurfactant was used to remove organic-bound copper and carbonate-bound zinc. Exchangeable, carbonate, reducible oxide, and organic fractions are amenable to washing techniques, and residually bound contaminants are not economical or feasible to remove. This information is important in designing the most appropriate conditions for sediment washing.

Although sequential extraction techniques work well for Cu, Ni, Co, Zn, and other metals, it is not appropriate for mercury. Other techniques have thus been devised. Bloom et al. (2003) used various reagents to define the behavior of mercury. The extracted Hg was defined as water-soluble Hg, stomach acid-soluble Hg, organo-chelated Hg, elemental Hg, and mercuric sulfide. Shi et al. (2005) studied sediment samples from the Haihe and Dagu Rivers in China. The elemental mercury and mercuric sulfide accounted for 46.5% and 39.0% of the total amount of mercury, which is considered to be not available. The moderately available organo-chelated mercury accounted for 13.3%. The very available fractions, the water soluble and acid soluble, made up 0.6 and 0.9% of the total mercury. Therefore, the total mercury analysis is not sufficient for evaluating the risk of the sediment.

To determine the speciation of metals in soils and sediments, various methods are used. Software such as PHREEQC from the United States Geological Survey (http://wwwbrr.cr.usgs.gov/projects/GWC_coupled/phreeqc/index.html) can be used to simulate metal speciation for integration with transport models.

3.3.3 ORGANIC CHEMICAL POLLUTANTS

In the case of organic chemicals, partitioning is indicated by an equilibrium partition coefficient k_{ow}, which is a coefficient describing the ratio of the concentration of a specific organic pollutant in other solvents to that in water. This coefficient k_{ow}, which relates the water solubility of an organic chemical with its n-octanol solubility, is more correctly referred to as the n-octanol–water partition coefficient. The distribution of organic chemical pollutants between sediment fractions and pore water is generally known as partitioning. By this, chemical pollutants are partitioned such that a portion of the pollutants in the pore water (aqueous phase) is removed from the aqueous phase.

The parameter k_d is strongly related to f_{oc} (fraction of organic carbon), and this relationship is often shown as $k_d = f_{oc} k_{oc}$, where k_{oc} is the sediment–water distribution coefficient. Although it is not always linear adsorption with soil/sediment organic matter (SOM), k_{oc} is important in fate and transport modeling. The sorption coefficients must be as site specific as possible to reflect the conditions of weathering and aerobic/anaerobic effects.

Estimations of k_{oc} have been through one-parameter linear free energy relationships over many decades. Quantitative structure–activity relationships (QSARs) have been developed, and correlations between $\log k_{oc}$ and $\log k_{ow}$ (the octanol–water partition coefficient) and between $\log k_{oc}$ and $\log S_w$ (water solubility) have also been utilized. Polyparameter linear free energy relationships for estimating k_{oc} have been developed because the one-parameter linear relationships are not accurate for polar chemicals (Nguyen et al., 2005). Various factors can influence the k_{oc}. For the sorption of oils, the concentration of the oil and the weathering state must be accounted for (Jonker et al., 2003).

Witt et al. (2002) showed that, at the Dover site where the organic fraction is low (f_{oc}=0.00025), R values for perchloroethylene (PCE), trichloroethylene (TCE), and dichloroethylene (DCE) were determined as 1.3, 1.2, and 1.1, respectively. At R=1.3, transport across the site would take about 49 years. Polycyclic aromatic hydrocarbons (PAHs) with increasing molecular weights exhibit higher low k_{ow} and are thus bound more strongly to organic matter.

The partitioning of organic chemical pollutants is a function of several kinds of interacting mechanisms between the organic chemicals and the sediment solids in the natural sediment–water system. A key factor in the development of the kinds of interaction mechanisms is the type or class of organic chemicals. The degree of water solubility of the organic chemical is a key element. Non-aqueous-phase liquids (NAPLs) include those that are denser and lighter than water. The DNAPLs (dense NAPLs) include the organohalides and oxygen-containing organic compounds, and

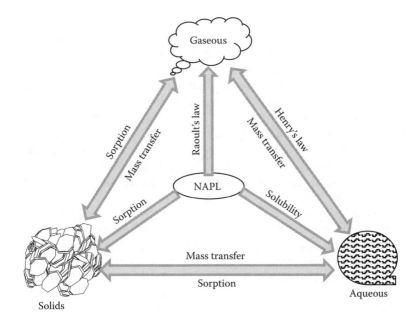

FIGURE 3.9 Processes involved in partitioning and fate of NAPLs.

vthe LNAPLs (light NAPLs) include gasoline, heating oil, kerosene, and aviation fuel. Most NAPLs are partially miscible in water.

The basic processes involved in the transport and fate of NAPLs are demonstrated in Figure 3.9. The chemical properties that affect NAPL transport and fate include (1) volatility, (2) relative polarity, (3) affinity for soil organic matter or organic contaminants, and (4) density and viscosity. The higher the vapor pressure of the substance, the more likely it is to evaporate. Movement in the vapor phase is generally by advection. At equilibrium between NAPLs and the vapor phase, the equilibrium partial pressure of a component is directly related to the mole fraction and the pure constituent vapor pressure as described by Raoult's law. Designating P_i as the partial pressure of the constituent, X_i as the mole fraction of the constituent, and P_i^0 as the vapor pressure of the pure constituent, Raoult's law states that, when equilibrium conditions are obtained, and when the mole fraction of a constituent is greater than 0.9, $P_i = X_i P_i^0$.

As shown in Figure 3.9, an organic chemical compound in the sediment may be partitioned between the pore water and the sediment constituents. The rate of volatilization of an organic molecule from an adsorption site on the solid phase in the sediment to the vapor phase in the air above the surface water is dependent on many physical and chemical properties of both the chemical and the soil, and on the process involved in moving from one phase to another. The two main distribution or transport processes involved are:

1. compound in sediment ↔ compound in solution
2. compound in solution ↔ compound in atmosphere

Partitioning of a chemical among these phases can be estimated from either vapor-phase or solution-phase desorption isotherms. The process by which a compound evaporates in the vapor phase to the atmosphere from another environmental compartment is defined as volatilization. This process is responsible for the loss of chemicals from the surface water to the air and is one of the factors involved in the persistence of an organic chemical. Determination of volatilization of a chemical from the sediment to the air is most often achieved using theoretical descriptions of the physical process of volatilization based on Raoult's law and Henry's law. The rate at which a chemical volatilizes from soil is affected by sediment and chemical properties and environmental conditions. Some of the properties of a chemical involved in volatilization are its vapor pressure, solubility in water, basic structural type, and the number, nature, and position of its basic functional groups.

Adsorption impacts directly on the chemical activity by reducing it to values below that of the pure compound. In turn, this affects the vapor density and the volatilization rate, because vapor density is directly related to the volatilization rate. Vapor density is the concentration of a chemical in the air, the maximum concentration being a saturated vapor. The role of water content is seen in terms of competition for adsorption sites on the soil. Displacement of nonpolar and weakly polar compounds by water molecules can occur because of preferential sorption (of water). Hydrates (i.e., hydration layer on the soil particle surfaces) will increase the vapor density of weakly polar compounds. If dehydration occurs, the compound sorbs onto the dry soil particles. This means that the chance for volatilization of the organic chemical compound is better when hydrates are present.

When a vapor is in equilibrium with its solution in some other solvent, the equilibrium partial pressure of a constituent is directly related to the mole fraction of the constituent in the aqueous phase. Once again, designating P_i as the partial pressure of the constituent, X_i as the mole fraction of the constituent in the aqueous phase, and H_i as Henry's constant for the constituent, Henry's law states that: $P_i = H_i X_i$. By and large, so long as the activity coefficients remain relatively constant, the concentrations of any single molecular species in two phases in equilibrium with each other will show a constant ratio to each other. This assumes ideal behavior in water and the absence of significant solute–solute interactions and also absence of strong specific solute–solvent interactions.

Partitioning of organic chemicals is most often described by the partition coefficient k_{ow}. This is the octanol–water partition coefficient and has been widely adopted in studies of the environmental fate of organic chemicals. The octanol–water partition coefficient is sometimes known as the equilibrium partition coefficient, which relates the ratio of the concentration of a specific organic pollutant in other solvents to that in water.

Results of countless studies have shown that this coefficient is well correlated to water solubilities of most organic chemicals. Since n-octanol is part lipophilic and part hydrophilic (i.e., it is amphiphilic), it has the capability to accommodate organic chemicals with the various kinds of functional groups. The dissolution of n-octanol in water is roughly eight octanol molecules to 100,000 water molecules in an aqueous phase. This represents a ratio of about one to twelve thousand (Schwarzenbach et al., 1993). Since water-saturated n-octanol has a molar volume of 0.121 L/mol as compared to 0.16 L/mol for pure *n*-octanol, the close similarity permits one to ignore the effect of the water volume on the molar volume of the organic phase in experiments conducted to determine the octanol–water equilibrium partition coefficient. The octanol–water partition coefficient k_{ow} has been found to be sufficiently correlated not only to water solubility, but also to soil sorption coefficients. In the experimental measurements reported, the octanol is considered to be the surrogate for soil organic matter.

Organic chemicals with k_{ow} values less than 10 are considered to be relatively hydrophilic—with high water solubilities and small soil adsorption coefficients. Organic chemicals with k_{ow} values greater than 10^4 are considered to be very hydrophobic and are not very water soluble. Chiou et al. (1982) has provided a relationship between k_{ow} and water solubility S as follows:

$$log\ k_{ow} = 4.5 - 0.75\ log\ S\ (ppm) \tag{3.6}$$

Aqueous concentrations of hydrophobic organics such as polyaromatic hydrocarbons, (PAH), in natural soil–water systems are highly dependent on adsorption/desorption equilibrium with sorbents present in the systems. Studies of compounds which included normal PAHs, nitrogen and sulfur heterocyclic PAHs, and some substituted aromatic compounds suggest that the sorption of hydrophobic molecules (with the exception of benzidine) is governed by the organic content of the substrate. The dominant mechanism of organic adsorption is the hydrophobic bond established between a chemical and natural organic matter in the sediment. The extent of sorption can be reasonably estimated if the organic carbon content of the sediment is known (Karickhoff, 1984) by using the expression: $k_d = k_{oc}\ f_{oc}$, where f_{oc} is the organic carbon content of the soil organic matter, k_{oc} is the organic content coefficient, and k_d is the linear Freundlich isotherm obtained for the target organic chemical. This approach works reasonably well in the case of high organic contents (e.g., $f_{oc} > 0.001$). Relationships reported in the literature relating k_{ow} to k_{oc} show that these can be grouped into certain types of organic chemicals.

For PAHs, the relationship given by Karickhoff et al. (1979) is:

$$log\ k_{oc} = log\ k_{ow} - 0.21 \tag{3.7}$$

For pesticides, Rao and Davidson (1980) report that:

$$log\ k_{oc} = 1.029\ log\ k_{ow} - 0.18 \tag{3.8}$$

For chlorinated and methylated benzenes, the relationship given by Schwarzenbach and Westall (1981) is:

$$log\ k_{oc} = 0.72\ log\ k_{ow} + 0.49 \tag{3.9}$$

The graphical relationship shown in Figure 3.10 uses some representative values reported in the various handbooks (e.g., Montgomery and Welkom, 1991; Verscheuren, 1983) for $log\ k_{ow}$ and $log\ k_{oc}$. The values used for $log\ k_{ow}$ are essentially

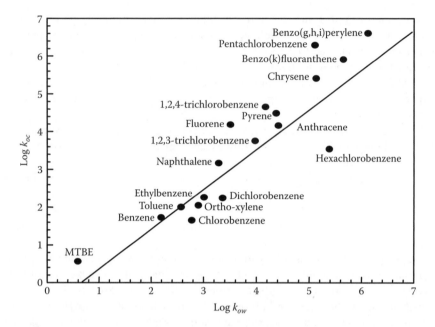

FIGURE 3.10   Relationship of log k_{oc} with log k_{ow} for several organic compounds (adapted from Yong and Mulligan, 2004).

mid-range results reported in the handbooks and in many studies. Not all *log k_{oc}* values are obtained as measured values. Many of these have been obtained through application of the various *log k_{oc}–log k_{ow}* relationships reported in the literature, such as those by Kenaga and Goring (1980) and Karickhoff et al. (1979). The linear relationship shown by the solid line in Figure 3.10 is given as:

$$log \; k_{oc} = 1.06 \; log \; k_{ow} - 0.68 \qquad (3.10)$$

This graphical relationship is useful in the sense of partitioning of the organic chemical compounds shown. Yong and Mulligan (2004) have discussed some of the pertinent correlations, stating, for example, in regard to the k_{oc} values shown in Figure 3.10 for dichlorobenzene, that they indicate that it partitions well to sediments, and particularly to the organic fractions (SOM, soil organic matter). Because of its resistance to anaerobic degradation, it is very persistent.

3.4   BIOTRANSFORMATION AND DEGRADATION OF ORGANIC CHEMICALS AND HEAVY METALS

The various types of organisms and microorganisms responsible for the biotransformation (this includes degradation) of inorganic and organic chemical compounds can be summarized as in Yong and Mulligan (2004) as follows.

Protozoa include pseudopods, flagellates, amoebas, ciliates, and parasitic protozoa. Their sizes usually vary from 10 to 50 μm, but can be up to 1 mm. They are aerobic, single-celled chemoheterotrophs and are eukaryotes with no cell walls.

Fungi are aerobic, multicellular eukaryotes and chemoheterotrophs that require organic compounds for energy and carbon. They reproduce by formation of asexual spores. In comparison to bacteria, they (a) do not require as much nitrogen, (b) are more sensitive to changes in moisture levels, (c) are larger, (d) grow more slowly, and (e) can grow in a more acidic pH range (less than pH 5). Fungi mainly live in the soil or on dead plants and are sometimes found in fresh water.

Algae are single-celled and multicellular microorganisms that are green, greenish tan to golden brown, yellow to golden brown (marine), or red (marine). They grow in the soil and on trees or in fresh or salt water. Those that grow with fungi are called lichens. Seaweeds and kelps are examples of algae. Since they are photosynthetic, they can produce oxygen, new cells from carbon dioxide or bicarbonate (HCO_3^-), and dissolved nutrients including nitrogen and phosphorus. They use light of wavelengths between 300 and 700 nm. Red tides are indicative of excessive growth of dinoflagellates in the sea. The green color in a body of lakes and rivers is eutrophication due to the accumulation of nutrients such as fertilizers in the water.

Although viruses are smaller than bacteria and require a living cell to reproduce, their relationship to other organisms is not clear. In order for them to replicate, they have to invade various kinds of cells. They consist of one strand of DNA or one strand of ribonucleic acid (RNA) within a protein coat. A virus can only attack a specific host. For example, those that attack bacteria are called bacteriophages.

The most significant animals in the soil are millimeter-sized worms. Nematodes are cylindrical in shape and are able to move within bacterial flocs. Flatworms such as tapeworms, eel worms, roundworms, and threadworms, which are nematodes, can cause diseases such as roundworm, hookworm, and filariasis.

Bacteria are prokaryotes that reproduce by binary fission by dividing into two cells, in about 20 minutes. The time it takes for one cell to double, however, depends on the temperature and species. For example, the optimal doubling time for *Bacillus subtilis* (37°C) is 24 minutes and for *Nitrobacter agilis* (27°C) is 20 hours. Classification is by shape, such as the rod-shaped bacillus, the spherical-shaped coccus, and the spiral-shaped spirillum. Rods usually have diameters of 0.5 to 1 micrometer and lengths of 3 to 5 micrometers. The diameter of spherical cells varies from 0.2 to 2 micrometers. Spiral-shaped cells range from 0.3 to 5 micrometers in diameter and 6 to 15 micrometers in length. The cells grow in clusters, chains, or in single form and may or may not be motile. The substrate of the bacteria must be soluble. In most cases, classification is according to the genus and species (e.g., *Pseudomonas aeruginosa* and *Bacillus subtilis*). Some of the most common species are *Pseudomonas, Arthrobacter, Bacillus, Acinetobacter, Micrococcus, Vibrio, Achromobacter, Brevibacterium, Flavobacterium,* and *Corynebacterium.* Within each species, there are various strains. Some strains that are better adapted can survive in certain conditions better than others. For survival, mutant strains can originate due to problems in the genetic copying mechanisms. Degradation of chemicals to an intermediate stage by one species of bacteria may be required for the growth of another species that utilizes the intermediate.

3.4.1 Bioremediation Processes

Understanding the types of chemicals that can be biodegraded or transformed, and the pathways of conversion, are important and will be discussed, as well as the toxicity and availability of several chemicals because this will serve as the foundation of knowledge required for determining the potential for natural attenuation. These concepts are described in further detail in Yong and Mulligan (2004).

Biodegradation is the prevalent mechanism for organic compounds which can be broken down into carbon dioxide and water. Some chlorinated compounds, however, may be more toxic than the parent compounds such as some lower-chlorinated dioxins (Safe, 1990). Aerobic degradation of polycyclic chlorinated biphenyls (PCBs) can occur (particularly those of four or fewer chlorines) (Unterman et al., 1987). In marine and freshwater sediment, dechlorination of meta- and para-chlorines is known. In Hudson River sediments, the half-lives for these types of chlorine were as little as three years (Brown et al., 1987). As PCBs dechlorinate, they become less toxic but more soluble, volatile, and bioavailable. However, there is some evidence that these by-products can accumulate and persist in the sediments and are thus more toxic (USEPA, 1996).

Polycyclic aromatic hydrocarbons (PAHs) degrade more quickly under aerobic than anaerobic conditions and thus tend to persist longer in anaerobic sediments (USEPA, 1996). There is low bioavailability for the microorganisms in the sediment, and they may also not have the enzymes to degrade the PAHs. Four- and five-ring PAHs are difficult to biodegrade compared to the smaller two- and three-ring PAHs.

Biodegradation rates in the literature should be used with caution in modeling studies for very numerous reasons. Rates can be one-tenth the initial levels after a year of weathering. Most data originates from spiked samples. Contaminated sediment may be anaerobic, and not aerobic where the rates are much faster. In addition, often contaminants are found as mixtures of many types. This can then interfere with the degradation of the compounds. Lower levels of the contaminants may also be limiting, slowing degradation. Therefore, there is significant uncertainty regarding biodegradation rates and whether the rates for recovery can be achieved in an acceptable time frame.

Microorganisms, the key to the biological treatment of contaminants, include bacteria, protozoa, fungi, algae, and viruses (Mulligan, 2002). A wide variety of hydrocarbons can be degraded by microorganisms through electron transfer by various mechanisms. Most of the knowledge related to natural attenuation is related to the degradation of BTEX compounds (benzene, toluene, ethylbenzene, and xylenes). The availability of oxygen and other electron acceptors such as nitrate, sulfate, and iron (III) determine the rate of biodegradation. However, anaerobic methanogenic degradation of benzene in aquifer sediments has been shown in the presence of water and mineral nutrients. Although the rate of aerobic biodegradation is higher than anaerobic, the latter type may be more dominant. Products of aerobic degradation are carbon dioxide and water, while the products of anaerobic degradation include carbon dioxide, water, methane, hydrogen, nitrogen, and others.

For chlorinated compounds, PCE and TCE, the reductive dehalogenation products of cis-DCE, vinyl chloride (VC), and ethane are indicators of degradation.

Highly chlorinated PCBs only partially degrade under anaerobic conditions (Bedard and May, 1996). Aerobic conditions are then needed to complete the degradation of the less-chlorinated congeners. However, the PCB-contaminated sediments can still exhibit toxicity and may pose a risk until completely treated (USEPA, 2005). Dechlorination rates in the natural environment range from seven to ten years for each chlorine removed.

The fuel oxygenate, methyl *tert*-butyl ether, MTBE, has been found in various surface water sources from atmospheric deposition, stormwater runoff, industrial releases, recreational activities, and discharge of groundwater. Microorganisms, however, within the stream and lake sediments are able to biodegrade the MTBE (Bradley et al., 2001). Increasing the silt and clay content decreased MTBE degradation almost completely, most likely due to a decrease in oxygen permeability into the sediments. Increasing the organic content also slightly correlated with a decrease in biodegradation. There may have been competition between the organic substrates.

Heavy metals are also subject to microbial conversions. Sediment samples, near a mining area in Spain of the largest producer of mercury (Hg), indicated concentrations of 2300 µg/g of Hg and 82 ng/g of methyl-Hg (Gray et al., 2004). These elevated methyl-Hg levels are an indication that microbial methylation is highly likely to occur in the wet, anoxic sediments with high organic contents, sulfate-reducing, or methanogenic conditions. Methyl-Hg is mobile, toxic, and bioaccumulates. Sulfate-reducing bacteria are important for mercury methylation, but have not been demonstrated in the field because sulfate concentrations in sea water and estuaries may be too high (Morel et al., 1998). HgS precipitation may also occur. Rittle et al. (1995) showed in the laboratory that arsenic could precipitate by bacterial sulfate reduction, thus immobilizing the arsenic on the sediments. Anaerobic conditions increase this mobility.

A uranium-contaminated sediment was obtained, and organic substrates were added (Suzuki et al., 2002). U(VI) can be reduced to U(IV) in the presence of organic substrates or electron acceptors such as Fe(III) oxides, sulfate, or selenate. The reduction removes uranium from solution. If in situ bioremediation was attempted in uranium mine ponds, *Desulfosporosinus* and *Clostridium spp.* would contribute to U(VI) reduction because they are commonly found in oxic sediments.

Sediment has also been obtained from a coal tailings pond (Siddique et al., 2007). *Enterobacter hormaechei*, a Se(VI)-reducing bacteria, and other Se(IV)-reducing bacteria were isolated from this sediment. They thus have the potential to precipitate the selenium in the aqueous phase.

Microcosm studies, polymerase chain reaction analysis (PCR), and site data can be used to determine the potential for natural attenuation at a site. Samples from both the groundwater and sediment are required. Microcosms are useful for identifying degradation potential under various nutrient and electron-donor conditions. For example, PCR analysis can provide information on the presence and spatial distribution of dechlorinating bacteria on site (Fennell et al., 2001).

Although the remediation of most sites is in temperate climates, the feasibility of natural attenuation at subarctic sites has been evaluated (Richmond et al., 2001). Although TCE and trichloroethane (TCA) degradation products were found, and reductive dechlorination conditions were likely, rates of degradation were slow, and thus dilution not biodegradation was the dominant attenuation mechanism. BTEX biodegradation was likely in the past. In situ sediment microcosm studies with organic acid measurement may be helpful in determining natural attenuation biodegradation mechanisms in dilute systems.

Various gases can be microbially produced in sediments, such as methane, hydrogen sulfide, and carbon dioxide. This can cause fissures in caps and desorb organic contaminants and thus must be considered in the long term. However, little information on gas generation has been obtained in caps for contaminated sediment management. Sulfate reduction in high-sulfate saline marine sediments can dominate. Other mechanisms include fermentation, denitrification, iron reduction, and methanogenesis.

3.4.2 BIOATTENUATION AND BIOAVAILABILITY

Determination of the capacity for bioattenuation has not received a great deal of attention in assessment of the natural attenuation of organic chemical pollutants. Substrates can become less bioavailable via interaction with negatively charged clay particles and organic material (Alexander, 1994). Sorption and sequestration can be influenced by pH, organic matter content, temperature, and pollutant characteristics. The biodegradation of PAHs is particularly affected by sorption.

Bioavailability of contaminants can influence microbial activity and biological responses. There exist many definitions of bioavailability, depending on the discipline. The National Research Council (National Research Council, 2002) in a recent report defined "bioavailability processes as the individual physical, chemical, and biological interactions that determine the exposure of organisms to chemicals associated with soils and sediments." Ehlers and Luthy (2003) recently attempted to define the terms bioaccessibility and bioavailability to improve risk assessment and remediation technology selection. Bioavailable contaminants are immediately available to an organism for storage, transformation, or biodegradation. This may not be sufficient for the microorganisms to be able to biodegrade the chemicals such as for PCBs. However, bioaccessible chemicals could be available to an organism after release from organic matter or other physical constraints after a short or lengthy time period. This is indicated in Figure 3.11.

Therefore, determination of sediment or soil contents could be used to indicate biodegradation potential. Changes in porosity can also occur as a result of dissolution processes. Excessive carbon dioxide produced can also increase porosity because of calcite and dolomite dissolution under acidic conditions (Bennett et al., 2000). Other reactions under anoxic conditions such as carbonate and bicarbonate saturation with calcite can plug pore spaces and decrease permeability.

An approach to screen for toxicity is to determine the acid volatile sulfide (AVS) ratio to simultaneously extracted metals (SEM). If AVS > SEM, then the divalent metals are not bioavailable or toxic. If the reverse is true, then the divalent metals may be bioavailable or toxic, and additional testing is required such as pore water analysis or toxicity characteristic leaching procedure (TCLP) (USEPA, 2005). To check bioavailability from the pore water characteristics, the following criteria can be followed.

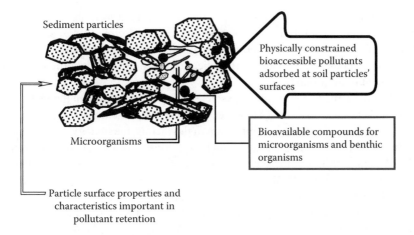

Sediment particles

Physically constrained
bioaccessible pollutants
adsorbed at soil particles'
surfaces

Microorganisms

Bioavailable compounds for
microorganisms and benthic
organisms

Particle surface properties and
characteristics important in
pollutant retention

FIGURE 3.11 Bioaccessible and bioavailable pollutants.

- Sum M/FCV < 1, then it is nontoxic.
- M = Metal (interstitial pore water), molar concentration of Cu, Pb, Ni, etc.
- FCV = final chronic value.

A variety of other products from bacteria can also influence the desorption of hydrocarbons and metals from the soil and sediments. Due to their anionic and hydrophilic/hydrophobic nature, biodegradable surfactants including rhamnolipids, surfactin, and sophorolipids, by-products of bacteria or yeast, have been able to remove metals and hydrocarbons from an oil-contaminated soil by disruption of the pollutant–sediment bonds (Mulligan, 2005).

3.5 INTERACTION OF CONTAMINANTS, ORGANISMS, AND SEDIMENTS

3.5.1 Bioaccumulation

Aquatic organisms can be in contact with or can ingest the sediments or suspended matter. Ingestion of sediments by bottom feeders will also have an influence on the contaminated sediments. The density of the freshwater oligochaetes can be up to 100,000 worms/m². They can then process between 10 to 20 times their own weight in the first 10 to 15 cm of sediments (Bentley et al., 2006). Tissue analyses or bioaccumulation studies can be used to study the accumulation of contaminants in the biota or food chain. However, the effect of the various contaminants is complicated. For example, although polycyclic aromatic hydrocarbons (PAHs) are commonly found in sediments, and many organisms can bioaccumulate PAHs into their tissues, fish do not accumulate PAHs, because they can metabolize and eliminate the PAHs. The PAHs, though, may reduce rates of growth and reproduction due to the toxicity of the metabolites. Lower invertebrate species may not as efficiently metabolize and eliminate PAHs. Due to sorption mechanisms, the fraction of the contaminant available to the organism may not be substantial for accumulation. Lu and Reible (2003) showed that bioaccumulation could be related to the contaminant concentration in the pore water. Thus accumulation can be an indication of the risk to benthic organisms by sediment contaminants.

Sulfate-reducing bacteria can form stable sulfide complexes with the metals cadmium, nickel, lead, and zinc (Ankley et al., 1996). When excess amounts of acid-volatile sulfides are present in comparison to the metals, the toxicity of these metals is reduced by precipitation, chelation, and sequestration of the metals (Berry et al., 1996). The availability of these metals for bioaccumulation in benthic invertebrates has been seen (Ankley et al., 1996). Redox reactions (with total organic carbon [TOC], Fe, or sulfide) may also reduce the solubility of metals, such as the conversion of Cr(VI) to the relatively stable form of Cr(III). The reduced form of Cr also has low mobility and toxicity.

The sulfides may also complex with various metals and inhibit methylation of metals such as mercury (Choi and Bartha, 1994). The methyl form of mercury is more toxic and bioaccumulative than inorganic mercury. The methylation process is an undesirable remediation process for monitored natural recovery (MNR). Microorganisms may also assist in the solubilization of organic matter and mineral phases that sequester the contaminants such as heavy metals. The mechanisms of mercury exchange between the water and sediments are not well understood.

As lipid solubility is characterized by k_{ow}, the k_{ow} of $HgCl_2$ was determined and found to be 3.3 (Morel et al., 1998). This indicates that the solubility in water and lipids is very similar. This compares to charged chloride complexes where the $k_{ow} = 0.5$. Therefore the uncharged form can diffuse much more quickly than the charged forms. The methyl form of mercury has a $k_{ow} = 1.7$, which is similar to the uncharged chloride form. The accumulation of methylmercury in the food chain is well known. Low pH and high concentration of chloride favor the accumulation.

3.5.2 BIOTURBATION

The displacement and mixing of sediment particles by benthic fauna or flora is called bioturbation. Various organisms involved in bioturbation include polychaetes, oligochaetes, bivalves such as mussels and clams, and gastropods. Faunal activities, such as burrowing, feeding, ingestion and defecation of sediment grains, and ventilation activities of benthic organisms are known as bioturbation. These activities displace sediment grains and mix the sediment matrix, which can affect chemical fluxes and thus exchange between the sediment and water column. Some organisms may further enhance chemical exchange by flushing their burrows with the overlying waters, a process termed bioirrigation. Benthic plants can affect sediments in a manner analogous to burrow construction and flushing by establishing root structures. Bioturbation is a diagenetic process that can change the physical and chemical structure of the sediment. Deposit feeders ingest sediment, move sediments toward the surface, and irrigate pore water. Sediment porosity and shear strength are also affected by bioturbation.

Modeling of these mechanisms is difficult due to a limited understanding. Banta and Andersen (2003) reviewed the mechanisms of the interaction of bioturbating organisms with sediment contaminants. Recalcitrant organic matter can be moved from anoxic to oxic zones, thus stimulating biodegradation. Insoluble metal complexes may also be oxidized which can then serve as electron acceptors for biodegradation. The focus of the review was on the polychaetes *Arenicola marina* and *Nereis diversicolor. A. marina* affects transportation via particle mixing, pore water flushing through irrigation, and degradation of organic pollutants via stimulation of microbial activity. *N. diversicolor* stimulates biodegradation directly through metabolism of the contaminants and affects biodiffusive mixing. It can also stimulate microbial activity. This study has indicated the complicated effects that bioturbation has on the fate and transport of contaminants in the sediment. Models should be mechanistically correct to predict the effect of bioturbation on the fate of pollutants as shown by the model by Forbes and Kure (1997). This model was coupled with an adsorption–degradation model by Timmermann (2001). Benthic organisms will also affect the biological processes within the sediments. In fresh water, they are mainly oligochaetes, with densities up to 100,000 worms/m² or more (Bentley et al., 2006). The organisms can process up to 10 to 20 times their body weight. The bioturbation zone and modeling simulation is shown in Figure 3.12.

The effect of the burrowing of the polychaete on sediment contaminated with 3,3′,4,4′-tetrachlorobiphenyl was evaluated (Gunnarsson et al., 1999). Bioturbation enhanced the release of the contaminant, and organic matter enhanced the release of the contaminant in the water column by 280% compared to the control. The enhanced release of the contaminant by the organic matter is contrary to other studies.

Grossi et al. (2002) evaluated the effect of benthic organisms on simulated oil spill-contaminated sediments. The organisms can rework the sediment, particularly at oxic/anoxic boundaries, which can have significant influence on microbial growth.

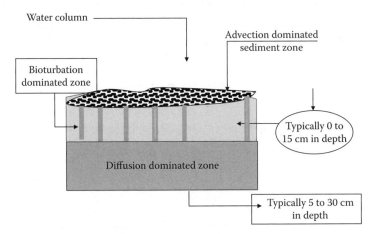

FIGURE 3.12 Schematic diagram showing contaminant transport processes within the sediment. Note there is some overlap of three regions. Cylindrical zones in the bioturbation zone represent mathematical simulation of bioirrigation zones (adapted from Burdige, 2006).

Acyclic compounds in particular are affected. It has been postulated that the digestive surfactants of the microfauna which ingest the sediment can enhance the solubility of the hydrocarbons and thus the biodegradation of the hydrocarbons.

Zinc fluxes in the presence of bioturbating organisms with and without capping were studied (Simpson et al., 2002). Without capping, zinc fluxes were on the order of 10 to 89 mg $Zn/m^2 \cdot day$. Removal of benthic organisms decreased bioturbation. Capping with clean sediment (5 mm thickness) was effective in reducing zinc fluxes by forming anoxic environments for the formation of metal sulfides. Capping materials were disturbed by the organisms, and therefore it was recommended that depths of the capping material must be greater than 30 cm.

3.6   CHEMICAL REACTIONS, GEOCHEMICAL SPECIATION, AND TRANSPORT PREDICTIONS

To meet the objectives of sustainability of the surface water environment, proper prediction of transport and fate of pollutants requires knowledge of how the abiotic and biotic reactions affect the long term health of the sediment system. From the various possibilities in handling the complex problem of chemical reactions and reaction rates and transformations, there exist at least four simple procedures that provide some accounting of the various processes controlling transport. These include (a) the addition of a reaction term r_c in the commonly used advection–diffusion equation given as Equation (3.11), (b) accounting for the contaminant adsorption–desorption process, (c) use of first- or second-order or higher-order reaction rates, and (d) combining transport models with geochemical speciation models. None of these appear to handle biotransformations and their resultant effect on the transport and fate processes.

Addition of a reaction term r_c to Equation (3.10) is perhaps the most common method used to accommodate a kinetic approach to fate and transport modeling. The resultant formulation is a linearly additive term to Equation (3.10) as follows;

$$R\frac{\partial c}{\partial t} = D_L \frac{\partial^2 c}{\partial x^2} - v \frac{\partial c}{\partial x} + r_c$$

(3.11)

The last term in Equation (3.11) can be expressed in the form of a general rate law as follows:

$$r_c = -k\vartheta |A|^a |B|^b$$

(3.12)

where r_c in this case is the rate of increase in concentration of a contaminant of species A, k is the rate coefficient, ϑ represents the volume of fluid under consideration, A and B are the reactant species, and a and b are the reaction orders.

The flux at the sediment–water interface can be given as:

$$\frac{N_A}{C_{pw} - C^*} = k_{sw} = \frac{1}{1/k_w + 1/(K_{sw}k_s\rho_s)} \tag{3.13}$$

where C_{pw} is the contaminant concentration in the pore water below the sediment surface, and C^* is the equilibrium pore water concentration. The mass transfer coefficients of k_w, k_s, and k_{sw} are for the water film, surface sediment layer, and sediment–water interface, respectively. The partition coefficient at the solid phase/pore water in the sediment is represented by K_{sw}. The average density of the particles is represented by ρ_s. Values for k_w have been determined to be in the ranges of 12.3 to 33.3 cm/day for rivers that have stable beds and are subject to erosion (Thibodeaux et al., 2002). The values for k_s were estimated by the same authors to be 2 to 3.6 cm/year. The term $(K_{sw}k_s\rho_s)$ can range from 5.5 to 1000 cm/day. The lower values correspond to low sorption and sediment reworking, while the higher end is associated with more reworking and more hydrophobic compounds. The water side thus controls the contaminant transfer in the latter case.

The abiotic reactions and transformations are sensitive to at least two factors: (a) the physicochemical properties of the pollutant itself and (b) the physicochemical properties of the sediment. Similar to inorganic contaminants, abiotic chemical reactions with organic compounds occur and include (a) hydrolysis, (b) formation of a double bond by removal of adjacent groups, and (c) oxidation–reduction, dehydrohalogenation, or hydrolysis reactions. For example, hydrolysis half-lives for PCE and TCE have been estimated as 9.9×10^8 and 1.3×10^6 years (Jeffers et al., 1989). In particular, anaerobic conditions promote the formation of metal sulfides of reduced solubility.

Abiotic reactions and transformations, together with the biotic counterparts, form the suite of processes that are involved in the transport and fate of contaminants in the sediment. The reactions between the chemical species in the pore water and also with the reactive sediment particle surfaces discussed in the previous sections and chapters constitute the basic platform. Because individual chemical species can participate in several types of reactions, the equations to describe the various equilibrium reactions can become complicated.

Geochemical modeling provides a useful means for handling the many kinds of calculations required to solve the various equilibrium reactions. Specific requirements are a robust thermodynamic database and simultaneous solution of the thermodynamic and mass balance equations. Appelo and Postma (1993) provide a comprehensive treatment of the various processes and reactions, together with a user guide for the geochemical model PHREEQE developed by Parkhurst et al. (1980). As with many of the popular models, the model is an aqueous model based upon ion-pairing and includes elements and both aqueous species and mineral phases (fractions).

Other available models include the commonly used MINTEQ (Felmy et al., 1984) and the more recent MINTEQA2 that includes PROFEFA2 (Allison et al., 1991), a preprocessing package for developing input files, GEOCHEM (Sposito and Mattigod, 1980), HYDROGEOCHEM (Yeh and Tripathi, 1990), and WATEQF (Plummer et al., 1976). By and large, most of the geochemical codes assume instantaneous equilibrium (i.e., kinetic reactions are not included in the calculations). In part, this is because reactions such as oxidation–reduction, precipitation–dissolution, substitution–hydrolysis, and to some extent, speciation–complexation, can be relatively slow. To overcome this, some of the models have been able to provide analyses that point toward possible trends and final equilibria. The code EQ6 (Delaney et al., 1986) does provide for consideration of dissolution–precipitation reactions. Transformations, however, are essentially not handled by most of the codes.

3.7 CONCLUDING REMARKS

Pollutants and contaminants and the manner in which they are handled in respect to the environment impact greatly on whether we can work toward sustainability of the environment or natural resources. The impact of these (pollutants and contaminants) and the implementation of indicators as a technique for assessment need proper consideration. Figure 3.13 summarizes some of the tests for evaluating the risk of the pollutants in the surface water–sediment environment.

In the assessment of the impact, an understanding of the interactions within the sediment is required to evaluate the goals of sustainability. The prediction of the transport and fate of pollutants is required. However, this is not a simple problem that can be handled with one set of tools. Analytical computer modeling is the most common technique used to provide information to predict system behavior. The limitations of the implementation of such models include the availability of appropriate and realistic input parametric information (especially partition and dis-

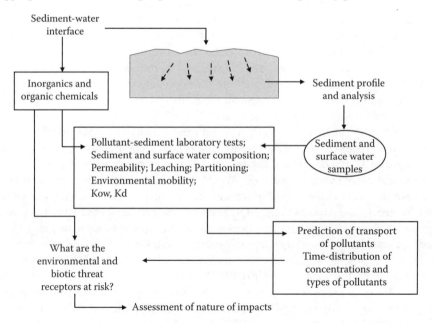

FIGURE 3.13   Schematic showing procedures, factors, tests and analyses required to begin the process for determination of consequence of pollutant discharge.

tribution coefficients) and chemical reactions that affect the status of the pollutants in the system.

The use of geochemical speciation modeling allows one to determine these reactions. However, since kinetic reactions are not readily handled in the available geochemical models, and since most of these models are not coupled to the regular transport models, much work remains at hand to obtain a reactive prediction model that can tell us about the fate and transport of pollutants in the surface water system.

REFERENCES

Alexander, M. 1994. *Biodegradation and Bioremediation*, Academic Press, San Diego.

Allison, J.D., Brown, D.S., and Novo-Gradac, K.J. 1991. MINTEQA2/PRODEFA2, a geochemical assessment model for environmental systems. USEPA, 1991.

Ankley, G.T., Di Toro, D.M., Hansen, D.J., and Berry, W.J. 1996. Technical basis and proposal for deriving sediment quality criteria for metals. *Environ. Toxicol. Chem.* 15: 2056–2066.

Appelo, C.A.J. and Postma, D. 1993. *Geochemistry, Groundwater and Pollution.* Balkema, Rotterdam, 536 p.

Banta, G. and Andersen, O. 2003. Bioturbation and the fate of sediment pollutants—Experimental case studies of selected infauna species. *Vie et Mileu* 53: 233–248.

Bedard, D.L. and May, R.J. 1996. Characterization of the polychlorinated biphenyls (PCBs) in the sediments of Woods Pond: Evidence for microbial dechlorination *in situ. Environ. Sci. Technol.* 30: 237–245.

Bennett, P.C, Hiebert, F.K., and Roger, J.R. 2000. Microbial control of mineral-groundwater equilibria: Macroscale to microscale. *Hydrogeology J.* 8: 47–92.

Bentley, S., Thibodeaux, L., Adriaens, P., Li, M.-Y., Romero-Gonzalez, M., Banwart, S.A., Filip, Z., Demnerova, K., and Reible, D. 2006. Physicochemical and biological assessment and characterization of contaminated sediments. In D. Reible and T. Lanczos, (Eds.), *Assessment and Remediation of Contaminated Sediments.* Springer, Dordrecht, Netherlands, pp. 83–136,

Berry, W.J., Hansen, D.H., Mahoney, J.D., Robson, D.L., Di Toro, D.M., Shipley, B.P., Rogers, B., Corbin, J.M., and Boothman, W.S. 1996. Predicting the toxicity of metal-spiked laboratory sediments using acid-volatile sulphide and pore water normalizations. *Environ. Toxicol. Chem.* 15: 2067–2079.

Bloom, N.S., Preus, E., Katon, J., and Hiltner, M. 2003. Selective extractions to assess the biogeochemically relevant fractionation of inorganic mercury in sediments. *Anal. Chim. Acta* 479: 233–248.

Bostick, B.C., Chen, C., and Fendorf, S. 2004. Arsenite retention mechanisms with estuarine sediments of Pescadero. CA, *Environ. Sci. Technol.* 35: 3299–3304.

Bradley, P.M., Landmeyer, J.E., and Chapelle, F.H. 2001. Widespread potential for microbial MTBE degradation in surface-water sediments. *Environ. Sci. Technol.* 35: 658–662.

Brown, J.F., Bedard, D.L., Brennan, M.J., Carnahan, J.C., Feng, H. and Wagner, R.E. 1987. Polychlorinated biphenyl dechlorination in aquatic sediments. *Science*, 236, 709–712.

Burdige, D.J. 2006. *Geochemistry of Marine Sediments.* Princeton University Press, Princeton, NJ.

Carey, G.R., van Geel, P.J., Murphy, J.R., McBean, E.A., and Rover, F.A. 1998. Full-scale field application of a coupled biodegradation-redox model BIOREDOX, In G.B Wickramanayake and R.H., Hinchee, (Eds.), *Natural Attenuation of Chlorinated Solvents.* Battelle Press, Columbus, OH, pp. 213–218.

Chartier, M., Mercier, G., and Blais, J.F. 2001. Partitioning of trace metals before and after biological removal of metals from sediments. *Water Resources* 35(6): 1435–1444.

Choi, S.-C. and Bartha, R. 1994. Environmental factors affecting mercury methylation in estuarine sediments. *Bull. Environ. Contam. Toxicol.* 53(6): 805–812.

Chiou, G.T., Schmedding, D.W., and Manes, M. 1982. Partition of organic compounds on octanol-water system. *Environ. Sci. Technol.* 16: 4–10.

Delaney, J.M., Puigdomenech, I., and Wolery, T.J. 1986. Precipitation kinetics option of the EQ6 geochemical reaction path code. Lawrence Livermore National Laboratory Report, UCRL-56342, Livermore, CA, 44 p.

Ehlers, L.J. and Luthy, R.G. 2003. Contaminant bioavailability in soil and sediment. *Environ. Sci. Technol.* 37: 295A–302A.

Felmy, A.R., Girvin, D.C., and Jeene, E.A. 1984. MINTEQ—A computer program for calculating aqueous geochemical equilibria. PB84-157148, EPA-600/3-84-032 (February).

Fennell, D.E., Carroll, A.B., Gossett, J.M., and Zinder, S.H. 2001. Assessment of indigenous reductive dechlorinating potential at a TCE-contaminated site using microcosms, polymerase chain reaction analysis and site data. *Environ. Sci. Technol.* 35: 1830–1839.

Forbes, T.L. and Kure, L.K. 1997. Linking structure and function in marine sedimentary and terrestrial soil ecosystems: Implications for extrapolation from the laboratory to the field. In N.M. Van Straalen and H. Lokke (eds.), *Ecological Risk Assessment of Contaminants in Soil.*, Chapman and Hall, London.

Gismera, M.J., Lacal, J., Da Silva, P., Garcia, R., Sevilla, M.T., and Procopio, J.R. 2004. Study of metal fractionation in river sediments. A comparison between kinetic and sequential extraction procedures. *Environ. Poll.* 127(2): 175–182.

Gray, J.E., Hines, M.E., Higueras, P.L., Adatto, I., and Lasorsa, B.K. 2004. Mercury speciation and microbial transformation in mine wastes, stream sediments, and surface waters at the Almaden Mining District. *Environ. Sci. Technol.* 38: 4285–4292.

Grossi, V., Massias, D., Stora, G., and Bertrand, J.-C. 2002. Burial, exportation and degradation of acyclic hydrocarbons following a simulated oil spill in bioturbated Mediterranean coastal sediments. *Chemosphere* 48: 847–954.

Gunnaersson, J.S., Hollertz, K., and Rosenberg, R. 1999. Effects of organic enrichment and burrowing activity of the polychaete *Nereis diversicolor* on the fate of tetrachlorobiphenyl in marine sediments. *Environ. Toxicol. Chem.* 18: 1149–1156.

Ho, M. D. and Evans, G. J. 2000. Sequential extraction of metal contaminated soils with radiochemical assessment of readsorption effects. *Environ. Sci. Technol.* 34: 1030–1335.

Jeffers, P.M., Ward, L.M., Woytowitch, L.M., and Wolfe, N.L. 1989. Homogenous hydrolysis rate constants for selected chlorinated methanes, ethanes, ethenes, and propanes. *Environ. Sci. Technol.* 23: 965–969.

Jonker, M.T.O., Sinke,. A. C., Brils, J.M., and Koelmans, A.A. 2003. Sorption of polycyclic aromatic hydrocarbons to oil contaminated sediment: unresolved complex? *Environ. Sci. Technol.* 37: 5197–5203.

Karickhoff, S.W. 1984. Organic pollutants sorption in aquatic system. *J. Hydraul. Eng.* 110: 707–735.

Karickhoff, S.W., Brown, D.S., and Scott, T.A. 1979. Sorption of hydrophobic pollutants on natural sediments, *Water Resources.* 13: 241–248.

Kenaga, E.E. and Goring, C.A.I. 1980. Relationship between water solubility, soil sorption, octanol-water partitioning and concentration of chemicals in biota. ASTM-STP 707, pp. 78–115.

Koeckritz, T., Thoming, J., Gleyzes, C., and Odegard, K. E. 2001. Simplification of a sequential extraction scheme to determine the mobilisable heavy metal pool in soil. *Acta Hydrochem. Hydrobiol.* 29: 197–205.

Krishnamurthy, J.S.R., Huang, P.M., Van Rees, K.C.J., Kozzak, L.M., and Rstad, H.P.W. 1995. Speciation of particulate-bound cadmium of soils and its bioavailability. *Analyst.* 120: 659–665.

Lake, D.L. 1987. Chemical speciation of heavy metals in sewage sludge and related matrices. In Lester, J.N. (Ed.), *Heavy Metals in Wastewater and Sludge Treatment Processes, I: Sources, Analysis, and Legislation.* CRC Press, Boca Raton, FL.

Lerman, A. 1979. *Geochemical Processes: Water and Sediment Environments.* John Wiley and Sons, New York, 481 p.

Lewis, G.N. 1923. *Valences and the Structure of Atoms and Molecules.* The Chemical Catalogue, New York.

Li, Y.H. and Gregory, S. 1974. Diffusion of ions in sea water and in deep-sea sediments. *Geochem. Cosmochim. Acta* 38: 603–714.

Lu, X.X. and Reible, D. 2003. Linking sediment exposure with effects: modelling techniques, organic availability and uptake. *Int. J. Sed. Res.* 18(2): 208–213.

Mason, R.P., Fitzgerald, W.F., and Morel, F.M.M. 1994. The biogeochemical cycling of elemental mercury: anthropogenic influences. *Geochim. Cosmochim. Acta* 58: 3191–3198.

Montgomery, J.H. and Welkom, L.M. 1991. *Groundwater Chemicals Desk Reference.* Lewis Publ., Ann Arbor, MI, 640 p.

Morel, F.M., Krapiel, A.M.L. and Amyot, M. 1998. The chemical cycle and bioaccumulation of mercury. *Ann. Rev. Ecol. Syst.* 29: 543–566.

Mulligan, C.N. 2002. *Environmental Biotreatment.* Government Institutes, Rockville, MD.

Mulligan, C.N. 2005. Environmental applications for biosurfactants. *Environ. Poll.* 133: 183–198.

Mulligan, C.N. and Dahr Azma, B. 2003. Use of selective sequential extraction for the remediation of contaminated sediments. In J. Locat, R. Galvez-Cloutier, R.C. Chaney, and K. Demars (Eds.), *Contaminated Sediments: Characterization, Evaluation, Mitigation/ Restoration, and Management Strategy Performance, ASTM STP 1442*, ASTM International, West Conshohocken, PA, pp. 208–223.

National Research Council. 2002. *Bioavailability of Contaminants in Soils and Sediments: Processes, Tools and Applications.* National Academies Press, Washington, DC.

Nernst, W. 1888. Zur Kinetik der in Losung befinlichen Korper, *Zeitschrift fur Physikalishe Chemie.* 2: 613–637.

Newell, C.J., McLeod, R.K., and Gonzales, J. 1996. BIOSCREEN Natural attenuation decision support systems, Report EPA/6000/R-96/087, August.

Nguyen, T.H., Goss, K.-U., and Ball, W.P. 2005. Polyparameter linear free energy relationships for estimating the equilibrium partition of organic compounds between water and the natural organic matter in soils and sediments. *Environ. Sci. Technol.* 39: 913–924.

Ogata, A. and Banks, R.B. 1961. A solution of the differential equation of longitudinal dispersion in porous media, U.S. Geological Survey Paper 411-A.

Parkhurst, D.L., Thorstenson, D.C., and Plummer, L.N. 1980. PHREEQE—A computer program for geochemical calculations, US Geological Survey Water Resources Investigation, 80-96, 210 p.

Perkins, T.K. and Johnston, O.C. 1963. A review of diffusion and dispersion in porous media. *J. Soc. Petrol. Eng.* 17: 70–83.

Peters, R.W. 1999. Chelant extraction of heavy metals from contaminated soils. *J. Haz. Mat.* 66: 151–210.

Petrozelli, G., Giudi, G., and Lubrano, L. 1983. *Proc. Int. Conf. Heavy Metals in the Environment,* p. 475.

Pignatello, J.J., Ferrandino, F.J., and Huang, L.Q. 1993. Elution of aged and freshly added herbicides from a soil. *Environ. Sci. Technol.* 27: 1563–1571.

Plummer, L.N., Jones, B.F., and Truesdell, A.H. 1976. WATEQF—A FORTRAN IV version of WATEQ, a computer code for calculating chemical equilibria of natural waters. U.S. Geological Survey Water Resources Investigation, 76-13, 61 p.

Rao, P.S.C. and Davidson, J.M. 1980. Estimation of pesticide retention and transformation parameters required in nonpoint source pollution models. In M.R. Overcash and J.M. Davidson (Eds.), *Environmental Impact of Nonpoint Source Pollution*. Ann Arbor Science, Ann Arbor, MI, pp. 23–27.

Richmond, S.A., Lindstrom, J.E., and Braddock, J.F. 2001. Assessment of natural attenuation of chlorinated aliphatics and BTEX in subarctic groundwater. *Environ. Sci. Technol.* 35: 4038–4045.

Rittle, K.A., Drever, J.I., and Colberg, P.J.S. 1995. Precipitation of arsenic during bacterial sulfate reduction. *Geomicrobiol. J.* 13: 1–11.

Safe, S. 1990. Polychlorinated biphenyls (PCBs), dibenzo-p-dioxins (PCDDs), dibenzofurans (PCDFs) and related compounds: environmental and mechanistic considerations which support the development of toxic equivalency factors (TEFs) *Crit. Rev. Toxicol.* 21, 51–88.

Sandoli, R.L., Ghiorse, W.C., and Madsen, E.L. 1996. Regulation of microbial phenanthrene mineralization in sediment samples by sorbent-sorbate contact time, inocula and gamma irradiation-induced sterilization artifacts. *Environ. Toxicol. Chem.* 15(11): 1901–1907.

Sayles, F.L. and Mangelsdorf, P.C., Jr. 1977. The equilibration of clay minerals with sweater: exchange reactions. *Geochim Cosmochim. Acta* 41: 951–960.

Schramel, O., Michalke, B., and Kettrup, A. 2000. Study of the copper distribution in contaminated soils of hop fields by single and sequential extraction procedures. *Sci. Total Environ.* 263: 11–22.

Schwarzenbach, R.P. and Westall, J. 1981. Transport of non-polar organic compounds from surface water to groundwater: Laboratory sorption studies. *Environ. Sci. Technol.,* 15(11): 1360–1367.

Schwarzenbach, R.P., Gschwend, P.M., and Imboden, D.M. 1993. *Environmental Organic Chemistry.* John Wiley and Sons, New York, 681 p.

Shi, J.-B., Liang, L.-N., Jiang, G.-B., and Jin, X.-L. 2005. The speciation and bioavailability of mercury in sediments of Haihe River, China. *Environ. Internat.* 31(3): 357–365.

Shuman, L.M. 1985. Fractionation method for soil micro elements. *Soil Science* 140: 11–22.

Siddique, T., Arocena, J.M., Thring, R.W., and Zhang, Y. 2007. Bacterial reduction of selenium in coal mine tailings pond sediment, *J. Environ. Qual.* 36: 621–627.

Simpson, S.L., Pryor, I.D., Mewburn, B.R., Batley, G.E., and Jolley. D. 2002. Considerations for capping metal-contaminated sediments in dynamic estuarine environments. *Environ. Sci. Technol.* 36: 3772–3778.

Sposito, G. and Mattigod, S.V. 1980. GEOCHEM: A computer program for the calculation of chemical equilibria in soil solutions and other natural water systems. Dept. of Soils and Environment Report, University of California, Riverside, 92 p.

Steinberg, S.M., Pignatello, J.J., and Sawhney, B.L. 1987. Persistence of 1,2-dibromoethane in soils: Entrapment in intraparticle micropores. *Environ. Sci. Technol.* 21: 1201–1208.

Sun, Y., Petersen, J.N., Clement, T.P., and Hooker, B.S. 1996. A monitoring computer model for simulating natural attenuation of chlorinated organics in saturated groundwater aquifers. *Proc. Symp. Natural Attenuation of Chlorinated Organic in Groundwater*. Dallas, TX, RPA/540/R-96/509.

Suzuki, Y., Kelly, S.D., Kemner, K.M., and Banfield, J.F. 2002. Microbial populations stimulated for hexavalent uranium sediment in uranium mine sediment. *Appl. Environ. Microbiol.* 69: 1337–1346.

Tessier, A., Campbell, P.G.C., and Bisson, M. 1979. Sequential extraction for the separation of particulate trace metals. *Analytical Chemistry* 51: 844–851.

Tessier, A., Campbell, P.G.C., and Bisson, M. 1982. Particulate trace speciation in stream sediments and relationships with grain size: Implication for geochemical exploration. *J. Geochem Explor.* 16-2: 77–104.

Thibodeaux, L.J., Reible, D.D., and Valsaraj, K.T. 2002. Non-particle resuspension chemical transport from stream bed. In T. Lipnick, R. Mason, M. Philips and C. Pittman (Eds.), *Chemicals in the Environment: Fate, Impacts and Remediation,* Chapter 7, ACS Symposium Series, 806.

Timmermann, K. 2001. Effect and fate of pyrene in bioturbated sediment development, verification and use of diagenetic models. M.Sc. Department of Life Science & Chemistry, Roskilde Univ. Roskilde.

Unterman, R. A. 1996. History of PCB biodegradation, In: *Bioremediation: Principles and Applications,* Crawford, R.L.; Crawford, D.L. (Eds.), Cambridge University Press, New York, pp. 209.

USEPA. 1996. Bioremediation of sediments contaminated with polyaromatic hydrocarbons, EPA Grant No. R819165-01-0. Investigators C.H. Ward and J.B. Hughes.

USEPA. 2005. Contaminated Sediment Remediation Guidance for Hazardous Waste Sites. EPA540-R-05-012, Office of Solid Waste and Emergency Response, December 2005.

Verscheuren, K. 1983. *Handbook of Environmental Data on Organic Chemicals,* (2nd ed.). Van Nostrand Reinhold, New York, 1310 p.

Wilcock, R.J., Corban, G.A., Northcott, G.L., Wilkins, A.L., and Langdon, A.G. 1996. Persistence of polycyclic aromatic compounds of different molecular size and water solubility in surficial sediment of intertidal sandflat. *Environ. Toxicol. Chem.* 15(5): 670–676.

Witt, M.E., Klecka, G.M., Lutz, E.J., Ei, T.A., Grosso, N.R., and Chapelle, F.H. 2002. Natural attenuation of chlorinated solvents at Area 6, Dover Air Force Base: groundwater biogeochemistry. *J. Contam. Hydrol.* 57: 61–80.

Yeh, G.T. and Tripathi, V.S. 1990. HYDROGEOCHEM, a coupled model of HYDROlogic transport and GEOCHEMical equilibria in reactive multicomponent systems, Oak Ridge National Laboratory, Oak Ridge, TN.

Yong, R.N. 2001. *Geoenvironmental Engineering: Contaminated Soils, Pollutant Fate, and Mitigation,* CRC Press, Boca Raton, FL, 307 p.

Yong, R.N., Bentley, S.P., Harris, C., and Yaacob, W.Z.W. 1999. Selective sequential extraction analysis (SSE) on estuarine alluvium soils. *Proc. 2nd BGS Conference on Geoenvironmental Engineering,* Thomas Telford, London, pp.118–126.

Yong, R.N. and Mulligan, C.N. 2002. The impact of clay microstructural features on the natural attenuation of contaminants. *Proc. Workshop on Clay Microstructure and Its Importance to Soil Behaviour,* Lund, Sweden.

Yong, R.N. and Mulligan, C.N. 2004. *Natural Attenuation of Contaminants in Soils.* Lewis Publ., Boca Raton, FL, 319 p.

Yong, R.N. and Warkentin, B.P. 1975. Elsevier Scientific Publishing Co., Amsterdam, 449 pp.

Zeien, H., and Brummer, G. W. 1989. Chemische extrakionen zur bestimmung von schwermetallbindungsformen. *Boden. Mitt. Dtsch. Bodenkd. Ges.* vol. 59-1.

4 Remediation Assessment, Sampling, and Monitoring

4.1 INTRODUCTION

Sediment quality is related to the quality of surface water. It is due to the serial mechanisms of the dissolution of organic matter and the exclusion of contaminants resulting from the consolidation of sediments or the leaching of contaminants. In addition, the food chain starting from the bottom is the most important mechanism for the contamination of aquatic life (Bright et al., 1995; McLachlan et al., 2001). Therefore, in order to make an appropriate assessment of sediments, the physical, chemical, and biological mechanisms have to be understood well. Since the mechanisms are natural and complex, there is the possibility that nonpredictable results can be obtained. Therefore, it is required for engineers to modify or take measures suited to the occasion.

As a first step, existing site information needs to be examined to determine sampling needs. The information can include previous physical, chemical, and biological test results regarding the state and extent of contamination. Previous monitoring procedures and results should also be identified. The sources of the information often will be from environmental authorities or engineering reports. The latter, however, may not be widely available. Sometimes research publications or university reports may contain significant information on the site characteristics.

Once the literature search has been performed and reviewed, gaps in the data can be identified, and plans for sampling can then be made. The sampling of sediments and water is, therefore, required to determine the variables affecting the mechanisms. Monitoring can be used to predict future trends or to evaluate the progress of the remediation work. The scale for sampling and monitoring will be dependent on the projects. In many cases, the contamination of surface water is a primary concern because it is more easily seen by the public.

Because most of the physical and chemical properties of sediments have to be determined by the laboratory tests, sampling is almost always needed. Therefore, monitoring of sediment properties can be achieved by tests on samples obtained from the sites. Thus, much effort is required for the monitoring of sediments, including sampling procedures from a boat.

71

4.2 CLEANUP GOALS AND BACKGROUND VALUES

There are various sediment quality guidelines provided by different organizations or associations in different states or countries (Appendix D). In the United States, the Environmental Protection Agency (EPA), National Oceanic and Atmospheric Administration (NOAA), and the state governments of Washington and Florida have issued sediment quality guidelines. The EPA has issued "Procedures for the derivation of equilibrium partitioning sediment benchmark (ESBs) for the protection of benthic organisms: dieldrin, endrin, metal mixtures and PAH mixtures." Through its National Status and Trends (NS&T) Program, the NOAA developed a guideline for use as informal, interpretive tools for the NS&T Program. Critical levels of contamination are difficult to determine due to the interaction and effects of the numerous contaminants. In addition, the influence of heredity is complex.

Basically there are two guidelines for hazardous substances in water and sediments. One is the background values, which are natural levels without any human influence. This value is the most conservative value and the most ideal level. In Canada, this level is called the "interim quality guideline." For sediment, it is called the Interim Sediment Quality Guideline, ISQG (Canadian Sediment Quality Guidelines, 2003). In general, the interim quality guidelines are different for fresh and marine sediments.

Environment Canada and the Ministère du Développement durable, de l'Environnement et des Parcs du Québec (2008) published a report on the St. Lawrence plan for sustainable development. In the report, to protect aquatic life, the Canadian Council of Ministers of the Environment (CCME) has derived two reference values for 30 substances in freshwater and marine sediments: a threshold effect level (TEL) and a probable effect level (PEL). These two values have been adopted for the assessment of sediment quality in Quebec, and three other levels were derived to define all of the intervention levels needed for sediment management in Quebec under a diversity of contexts. The three new sediment quality criteria were defined using the CCME database and a calculation method similar to the one used to determine the TEL and the PEL. They are (1) the rare effect level (REL), (2) the occasional effect level (OEL), and (3) the frequent effect level (FEL). The guideline provides criteria for 8 metals and metalloids, and 26 organic compounds for freshwater and marine sediments, respectively (see Appendix A). According to the criteria, cleanup goals may be at the most the PEL or FEL, depending on the type of toxicity and other factors.

4.3 SAMPLING

A monitoring plan for sediments can be developed by the following steps:

1. monitoring item(s) and terms
2. selection of sampling methods, analytical instruments, devices, and/or techniques
3. site selection and date(s)
4. permission from the authorities concerned
5. chartered boat use, if necessary

Monitoring items may include the physical and/or chemical properties of the sediments. Various guidelines are given in Environment Canada (2002a,b). The selection of items depends on the objectives of monitoring. If the primary objective is to determine the contamination level of the sediments, chemical analyses will dominate. Therefore, sampling must follow the procedures provided by the appropriate standard or guideline. The standard may provide procedures for sampling, the types of samplers and containers, and storage methods prior to testing, etc. Contamination of the samples during the sampling must be avoided. The sampler and container must not react with or leach contaminants into the samples. Therefore, stainless steel and glass containers are preferred, while steel and aluminum materials are usually avoided. For metals and inorganic contaminants, high-density polyethylene (HDPE) or polytetrafluoroethlene (PFTE) are used for storage of sediments, while glass containers with PFTE lids are used for storing organic contaminated sediments. The glass containers should be cleaned using acetone.

The period of regular monitoring of sediment quality may be extended and infrequent, because sedimentation is usually slow and the situation changes slowly. Local and provincial or state governments in many countries have investigated sediment quality every year in rivers, lakes, port, and bays, in order to monitor the change in sediment quality. The frequency of investigation may be between once a month and once a year. Irregular monitoring of sediment quality can be made when contamination of the sediments is found, or when contaminated sediment was remediated and the result needs to be verified.

The sample obtained must be enough for sufficient analysis, from appropriate locations to characterize the contamination properly and accurately and enable identification of background values. The amount of sample collected will depend on the amount of analyses to be performed and the detection limit of the analyses. Large samples, however, may be difficult to transport. Sampling plans can be either random or biased. Replication of the samples decreases uncertainty but increases costs, so a balance must be obtained between the two. Collecting composite samples is frequently performed. Several samples can be combined as one, thus reducing costs for analysis. This procedure is more appropriate for homogeneous samples, because hot spots could be missed at heterogeneous sites.

Sediment samples can be obtained using a sampler. There are many types of samplers, as shown in Table 4.1. Core boring can be used to obtain mechanically undisturbed sediment samples. With this method, long core samples can be obtained that will provide an understanding of the deeper layers of the sediment. The core must be obtained from the vertical position. Costs can be high when the depth of the water is extensive.

The sampling method can be selected according to the following key factors:

1. Whether the water is shallow or deep
2. Whether the sample is undisturbed or disturbed
3. Whether it is a surface or core sample
4. Whether the sediment samples include benthos or not
5. The amount of sediments required

TABLE 4.1

Samplers and Analyses

Type of Sampler	Mechanical Properties	Physical Properties	Chemical Properties	Biological Properties	Comments
Boring	O	O	O	N	
Gravitational corer	P	O	O	N	A few meters
Grab samplers *Smith-McIntyre Ekman-Birge*	N	O	O	O	Surface only (a few centimeters)
Box sampler	P	O	O	O	Shallow

Note. O: OK, P: possible, N: Not possible.

Figure 4.1 shows an open core sampler with a square section and without a piston. The sampler has a cross-sectional area of 100 cm² and a length of 2 m. The advantages in the use of this sampler are that the sample taken can be observed immediately after sampling, by opening the side wall of the sampler, and that physical and simple mechanical tests such as vane shear and water content tests can be performed as shown in Figure 4.1. The disadvantage may be that sandy sediments cannot be obtained. The weight of the driving force is about 1000 N.

An improved sampler is a piston core sampler with a catcher (ASTM, 2006). A piston core sampler can be conveniently used even for deep water, as shown in Figure 4.2. The piston retains the core sample by the suction developed during the descent. The length of the core may be limited to several meters. A core diameter commonly used

FIGURE 4.1   Open core sampler and core sample obtained.

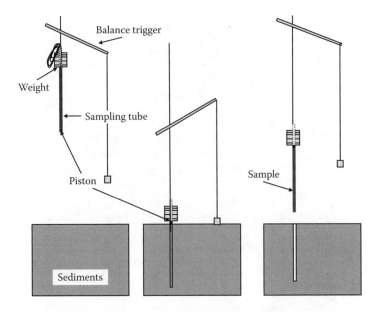

FIGURE 4.2 Procedures of sampling using piston core sampler.

is approximately 10 cm. The load (weight) for driven force is about 3000 N. The disadvantage of this method is that a large boat with a winch is required, and that sandy samples cannot be obtained without a catcher. A core sampler without a piston can also be used to obtain core samples. Variations can occur during core sampling.

Grab samplers are usually used to obtain surface and recently deposited sediments. The Smith-McIntyre grab sampler (Figure 4.3) is one of the most popular.

FIGURE 4.3 Smith-McIntyre grab sampler.

FIGURE 4.4   Birge-Ekman sampler.

This sampler needs a boat with a winch to remove it from the water. With this type, benthos can also be obtained. Therefore, the sampler is suitable for biological (benthos) sampling (ASTM, 2006). A smaller type of grab sampler, such as the Birge-Ekman sampler, can conveniently be used without mechanical assistance (Figure 4.4). However, because the sampler is very light, sandy samples cannot be easily obtained. There are various samplers available for collecting benthos (ASTM, 2006). The types and features of various samplers are summarized in Table 4.1.

Selection of the site can be made based on the history of contamination and the following considerations:

1. Representative site locations in a zone or area
2. A sufficient number of sampling points
3. Inclusion of sampling points in uncontaminated zones

When sampling includes the use of a boat, weather and sea conditions should be of the highest priority. Therefore, the scheduled dates should include auxiliary dates. All safety measures including the use of life jackets and hard hats should be followed.

All required document permissions should be obtained from the appropriate authorities. The procedures may vary in different countries. Early planning for boat chartering is optimal. It should also be remembered that a technician for the winch is also needed, if a heavy sampler is used.

4.4   ANALYSIS AND EVALUATION

While some measurements need to be very accurate, others do not need to be. To perform chemical analyses, an appropriate instrument should be used. However, a desired instrument may not always be available. In this case, accuracy and precision should be taken into account. If it is not satisfactory, the samples must be sent to external laboratories that are appropriately equipped.

4.4.1   MECHANICAL PROPERTIES

Mechanical properties of undisturbed sediments include strength and consolidation properties. The permeability of sediments can be included in this category, because it is an index of consolidation properties.

4.4.1.1   Strength for Sediments

The strength of sediments can be expressed as unconfined compressive strength or shear strength. The shear strength of sediments can be measured by the vane shear, unconfined compression, direct shear, or triaxial shear tests. These tests are very common in geotechnical engineering, and the tests methods are well established worldwide. However, surface sediments are usually so soft that the test specimen cannot be prepared easily. Accordingly, it is likely that only vane shear test is available for very soft sediments. There are two types of vane shear tests (i.e., in situ and laboratory tests). The in situ vane shear test may require a platform for testing and thus is usually avoided.

The laboratory vane shear test is a useful method of measuring the shear strength of clay because it is simple and quick. The laboratory vane shear test for the measurement of shear strength of cohesive soils is useful for soils of low shear strength (less than 30 kPa) for which triaxial or unconfined tests cannot be performed. The test gives the undrained shear strength of the soil. The undisturbed and remolded strengths obtained are also useful for evaluating the sensitivity of soil.

The vane itself is usually a four-bladed paddle that, when inserted into a sample and rotated slowly, causes the sample to deform and shear, as shown in Figure 4.5. The two cases provide different formulas for the shear strength, τ_f.

In the case of Figure 4.5(a), torque T is equal to $Pa/2$ and is given by

$$T = \tau_f \pi \left(\frac{d^2 h}{2} + \frac{d^3}{6} \right)$$

(4.1)

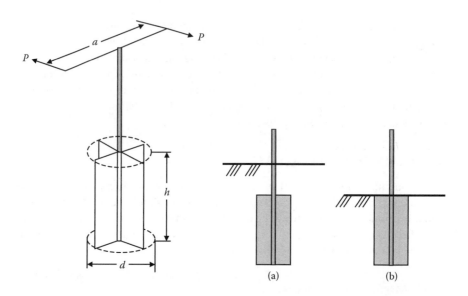

FIGURE 4.5 Vane shear test.

where T is the applied torque, d and h are the diameter and height, respectively, and τ_f is the vane shear strength of the sediment sample. From Equation (4.1), the shear strength is expressed by:

$$\tau_f = \frac{T}{\pi\left(\dfrac{d^2 h}{2} + \dfrac{d^3}{6}\right)} \qquad (4.2)$$

In the case of Figure 4.5(b),

$$\tau_f = \frac{T}{\pi\left(\dfrac{d^2 h}{2} + \dfrac{d^3}{12}\right)} \qquad (4.3)$$

Figure 4.6 shows the profile of vane shear strength on core samples obtained from Osaka Bay. The core sample was obtained with a piston core sampler. The calcium carbonate content was also obtained, and a strong correlation between vane shear strength and calcium carbonate content was found (Fukue et al., 1999), as can be seen in Figure 4.6.

From the analysis, it was found that the shear strength of the surface sediments increased by 7.5 kPa if the calcium carbonate content increased by 1%. This results from the cementation due to calcium carbonate (Fukue et al., 1999). A general trend shows that the increasing rate of shear strength with depth is 1.3 kPa/m, as shown in Figure 4.6. However, it includes the effect of the calcium carbonate content (i.e., 0.2%/m).

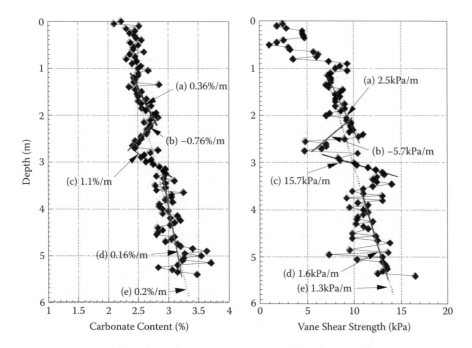

FIGURE 4.6   Vane shear strength and carbonate contents for shallow sediments in Osaka Bay.

The strength characteristics are important when the stability of bottom sediments is analyzed. For example, this can include instabilities due to the excavation of sediments or a cap load on the bottom.

From an environmental point of view, carbonates can retain metals in the sediments. A rough estimate using the results of the sequential extraction test and carbonate content shows that carbonate can retain a zinc content of 12.6 mg/g and lead of 1.9 mg/g. This value is quite high in comparison to other components, such as inorganic minerals and organic matter. In general, the concentrations of trace metals in ocean sediments are based on a carbonate-free basis (CFB).

4.4.1.2   Consolidation

The consolidation properties are used to predict settlement of sediment layer under a load, such as a sand cap. The settlement can occur due to the drainage of pore water from the sediment. Therefore, if the sediment is contaminated, contaminated pore water can be released from the sediment into the water column. This will be discussed further in Section 6.2.

4.4.2   Physical Properties

4.4.2.1   Sediment Temperature

Sediment temperature is the one of the most basic physical variables which changes seasonally. Since in situ measurement under water is not easy, sediment sample temperature should be measured immediately after it was obtained. Water depth should also be recorded. As the temperature of sediments increases, the degradation rate increases. Consequently, the dissolved oxygen in sediments is depleted, and the sediments will become anaerobic.

4.4.2.2 Grain Size

Grain size is one of the most important physical properties, because it is an index of specific surface area and can give information if the sediments are cohesive, granular, or in between. Classically, the grain size of sediment samples is determined by the sieve method for the coarse fractions, greater than 0.075 mm, and by the hydrometer method, based on the "Stokes" sedimentation rates, for the fine fractions. On the other hand, the laser diffraction size analysis, which is based on the forward scattering of monochromatic coherent light, is superior. If the amount of sample obtained is not enough for the classical grain size analyses, a laser diffraction size analysis can be used. There are basically two types of analyzers. One measures the diffraction of light, while the other utilizes the permeation of light (Furukawa et al., 2001).

An example of grain size distribution is shown in Figure 4.7. The curve is called grain size distribution curve, which indicates the percentage of particles finer than a particular size. In general, grain sizes of 10%, 30%, and 60% finer are described as D_{10}, D_{30}, and D_{60}, respectively, and are used to characterize the sediment, as follows.

$$\text{Coefficient of Uniformity, Uc} = \frac{D_{60}}{D_{10}} \tag{4.4}$$

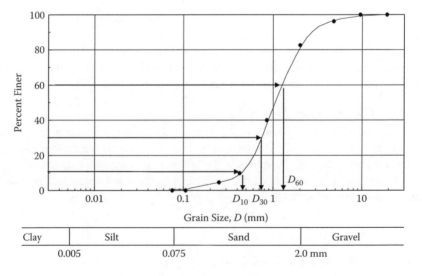

FIGURE 4.7   Example of a grain size distribution curve.

$$\text{Coefficient of Curvature, } Cc = \frac{D_{30}^{2}}{D_{10}D_{60}} \tag{4.5}$$

Sediment particles are defined according to size, as mentioned in Chapter 2. The smallest particle group which is less than 0.005 mm is called "clay." Particles from 0.005 mm to 0.075 mm belong to the "silt" category. The smallest size of "sand" is therefore 0.075 mm, and the maximum size is 2 mm. Particles greater than 2 mm are called "gravel."

The grain size of sediment particles can reflect the specific surface area (as discussed in Chapter 3), which strongly controls the boundary phenomena at the surface of particles. Clay particles have much greater specific surface areas than silt, sand, and gravel particles. Similarly, silt particles have much greater specific surface areas than sands and gravels (Fukue et al., 2006b).

Sediments are often classified as silty sand, clayey silt, silty clay, etc. These terms mean that sediments contain two or more fractions. The dominant component is named last. Measurement is by wet sieving, hydrometer testing, or laser particle size analysis. Sieves of large to small size are used to determine the amount of sediment that is retained on each sieve. The amount retained on each is determined by drying and weighing. Sieves usually range from 75 μm to 300 mm (JIS Z 8801), but this varies in different countries.

4.4.2.3 Specific Gravity

The specific gravity of particles, G_s, is defined as

$$G_s = \frac{\rho_s}{\rho_w} \tag{4.6}$$

where ρ_s and ρ_w are the densities of solid particles and pure water, respectively. The specific gravity of solid particles is determined by measuring the solid density with a pycnometer or a volumetric flask. The density of the particles at $T(°C)$ is calculated by

$$\rho_s = \frac{m_s}{m_s + (m_a - m_b)} \times \rho_w(T) \tag{4.7}$$

where

$$m_a = \frac{\rho_w(T)}{\rho_w(T')} \times (m_a{}' - m_f) + m_f \tag{4.8}$$

and m_s is the dry weight of particles, $m_a{}'$ is the weight of the pycnometer with distilled water at $T'(°C)$, m_b is the weight of the pycnometer involving particles and distilled water at $T(°C)$, and m_f is the weight of the pycnometer.

Primary minerals and secondary minerals have specific gravities ranging from 2.6 to 2.75, although the value always involves a small amount of organic matter and heavy minerals such as iron. The larger the amount of organic matter, the lower the specific gravity of sediment is. The specific gravity of organic-rich sediments may be as low as 1.5. Sediments of high density will settle faster. Laboratory tests can be performed to determine the settling characteristics of a sediment slurry. Turbidity measurements over time are performed. These types of measurements are particularly important to determine the potential effect of dredging.

4.4.3   Chemical Sediment Quality

The sediment quality variables depend on the purpose of the monitoring. Different countries have used different approaches and methods to describe and to measure sediment quality. The variables are grouped as shown in Figure 4.8.

4.4.3.1   pH

The pH is an important sediment measurement, which is often measured both at the sampling site and in the lab. pH meters are used for the measurements. Portable models are available to take out into the field. In general, calibration is required before measurement, using one or more solutions of known pH, such as buffer solutions with pH of 7.0 and 4.0.

Other parameters that should be measured on site will include redox potential, dissolved oxygen, conductivity/salinity, and turbidity. Most are measured with portable electrodes or probes.

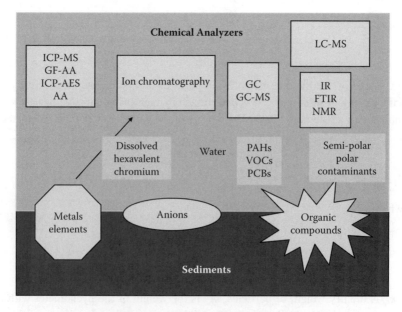

FIGURE 4.8   Grouping of environmental variables to be monitored.

4.4.3.2 Organic Pollution Indicators

In many countries, chemical oxygen demand (COD) is used as a dominant index for water quality which is related to the organic matter content in sediments. The COD is a measure of the oxygen requirement of a sample for oxidation by a strong chemical oxidant. The COD is determined using the analytical method (USEPA Method 410.4) or a COD meter (colorimetric method). The COD value for sediments is given by mg/g, and for water it is in mg/L. Higher COD values will cause adverse effects to water and sediment quality due to oxygen depletion. For water, COD is divided into particulate (PCOD) and dissolved (DCOD) fractions.

The Japan Fisheries Resources Conservation Association (2005) published the 2005 revised water quality for fisheries. The standard provided that COD value for sediments should be lower than 20 mg/g. In comparison to the standard, the actual measured values are considerably higher. For example, Thompson et al. (2001) obtained COD values greater than 300 mg/g for St. Lucie Estuary sediments with a volatile solid content ranging between 23% and 39%. However, this value may not be uncommon for anaerobic sediments with a high organic content.

4.4.3.3 Total Organic Carbon (TOC)

The total carbon (TC) in sediments consists of the total organic carbon (TOC) and the total inorganic carbon (TIC). Total organic carbon is separated into purgeable (POC) and nonpurgeable (NPOC) organic carbon. The NPOC is partitioned into particulate and dissolved (DOC) organic carbon.

TOC is the summation of all organic carbon compounds in water and is widely used. This is a monitoring parameter analyzed in environmental investigation programs. It is a chemical sediment factor that can influence the concentration of other compounds. Organic matter plays a major role in aquatic systems. It affects biogeochemical processes, nutrient cycling, biological availability, and chemical transport and interactions. It also has direct implications in the planning of wastewater treatment and drinking water treatment. The organic matter content is typically measured as TOC and DOC, which are essential components of the carbon cycle. Analysis is usually performed with CHN (carbon, hydrogen, and nitrogen) or TOC analyzers. Because many compounds can be bound to DOC such as polycyclic aromatic hydrocarbons (PAHs) and polycyclic chlorinated biphenyls (PCBs), it is an important parameter.

4.4.3.4 Loss on Ignition (Ignition Loss)

The organic matter content is often represented by loss on ignition (LOI or ignition loss). Ignition loss is defined as the percent of weight of volatile organic solids at a high temperature. The temperature used depends on the standards provided by different organizations. The lowest temperature may be 350°C and up to 850°C as a maximum.

Figure 4.9 shows a correlation between loss on ignition and TOC for soils and sediments in Japan. The correlation may depend on the types of organic matter at the site. These values are indicators of the content of organic matter and have been used to describe the sediment quality in both environmental and geotechnical points of views. In general, the lower the loss of ignition or TOC value, the better the quality.

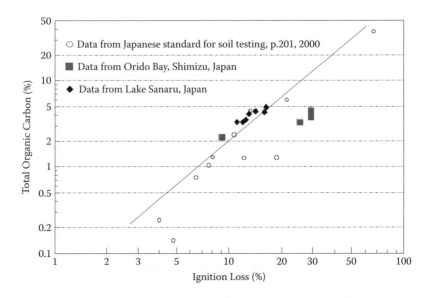

FIGURE 4.9   Correlation between TOC and ignition loss for various soils and sediments.

The organic matter will be degraded by microorganisms, depleting the oxygen in the water. If the loss on ignition is high, the water content of sediment is usually high. However, from an agricultural point of view, a higher content of organic matter is desirable, because of high contents of nutrients and water retention.

4.4.3.5   Nitrogen

Nitrogen is found in sediments and water in many forms, including inorganic, organic, dissolved, and particulate. Total nitrogen (abbreviated T-N) is a measure of all forms of dissolved and particulate nitrogen present in a sediment sample. Dissolved *inorganic* nitrogen is found as nitrate (NO_3^-), nitrite (NO_2^-), ammonium (NH_4^+), ammonia (NH_3), and nitrogen gas (N_2). Dissolved *organic* nitrogen is found in a wide range of chemical forms such as amino acids, proteins, urea, and humic acids. Total dissolved nitrogen consists of dissolved inorganic nitrogen and dissolved organic nitrogen and is readily available for plant uptake. The ammonium ion is the most readily available form of nitrogen available to phytoplankton.

The particulate nitrogen pool consists of plants and animals, their remains, and any ammonia attached to mineral particles. Particulate nitrogen can be found in suspension or in the sediment and is biologically available. T-N is a measure of all forms of dissolved and particulate nitrogen present in a water sample. Therefore, the T-N includes total Kjeldahl nitrogen (TKN), nitrate, and nitrite. TKN is the amount of organic nitrogen, and ammonia nitrogen (NH_4^+-N or NH_3-N). Therefore, TKN is a parameter that is often used as an indicator of industrial pollution and sewage.

A U.S. national soil nitrogen map which has been expressed by total Kejeldahl soil nitrogen at 1 km resolution was developed from the May 1994 National Soil Characterization Database, linked to the spatial information in STATSGO using soil taxonomic relationships. The TKN for soil is high in the deep Mollisols of the Midwest and in the Pacific Northwest (http://research.esd.ornl.gov/~hnw/esri98/).

For measurement, air-dried samples taken with grab samplers are used. There are many types of total nitrogen detectors. Some are portable, such as ammonium selective electrodes, and others are automatic laboratory analyzers. Colorimetric analyses are also widely used to determine various nitrogen forms.

Since nitrogen is abundant in the environment, and is not easily adsorbed on materials, the management and control of nitrogen in nature is difficult. Therefore, phosphorus has been managed or controlled for the measures of eutrophication. Nitrogen is released by denitrification from sediment into water or air. Denitrification is well known due to the action of bacteria such as *Pseudomonas* in paddy fields. This action can be applied for the removal of nitrogen from sediments.

4.4.3.6 Phosphorus

Phosphorus is an essential element necessary for growth of plants and animals, and high levels of phosphorus can be an important contributor to eutrophication, especially in lakes and marine estuarine systems. Eutrophication may cause algal blooms and seagrass decline. Excessive nutrient levels can cause odor, aesthetic problems, and fish kills. However, for reuse of dredged sediments for vegetation growth, nutrient content is important.

Phosphorus in aquatic systems is usually partitioned into particulate matter (organic and sediment) and dissolved fractions. Particulate phosphorus enters runoffs primarily through riparian litter fall, soil erosion, and sediment transport. They may exist in solution, as particles, loose fragments, or in the bodies of aquatic organisms. Measurement is expressed in mg/g. The total phosphorus analysis uses the thermal decomposition molybdate method or the photolytic decomposition molybdate method, among others.

4.4.3.6.1 Thermal Decomposition Method

A sample is decomposed by potassium peroxodisulfate and sodium hydroxide and then is cooled at an appropriate temperature. Ammonium molybdate is added to the cooled sample. This sample is then reduced with L-ascorbic acid to produce molybdate blue. The absorbance of the molybdate blue is obtained using a wavelength of 880 nm, and the concentration of total phosphorus (T-P) is measured. This method corresponds to the official method used for manual analysis.

4.4.3.6.2 Photolytic Decomposition Method

Potassium peroxodisulfate is added to the sample. The sample is irradiated with ultraviolet rays at approximately 95°C to decompose the sample through oxidation. Phosphoric compounds are then decomposed into phosphate ions. L-Ascorbic acid is added to this solution, and a zero calibration is performed. Sulfuric acid ammonium molybdate is added to color the solution, and then the absorbance is measured using a wavelength of 880 nm. The concentration of T-P in the sample water is measured. In this method, ultraviolet rays are radiated to provide the same effect as thermal decomposition at 120°C. In addition, operations are performed at normal pressures. There are many types of T-N/T-P detectors. Some are portable, and others are remote control analyzers. Most analyzers use absorption spectrophotometry. Chemical analyzers are also widely used to determine various phosphorus forms.

4.4.3.7   Toxic Substances—Trace Metals

For sediments and tissues, inductively coupled plasma mass spectrometry (ICP-MS), ICP-atomic emission spectrometry (AES), and flame atomic absorption spectrometry (AAS) techniques can be used to measure the total concentrations of metal elements (As, Cr, Cu, Cd, Pb, Hg, Ni, and Zn) in sediments, as shown in Figure 4.8, which shows chemical analyzers being used. Figure 4.10 shows an ICP-AES. The toxicity characteristic leaching procedure (TCLP) is used to evaluate the leachability of the treated and undertreated sediment. The extracts are analyzed for metal content and compared to the toxicity guidelines of the EPA to determine if the sediment is considered as a hazardous waste.

In many cases, sediments found at the bottom of the surface water are objects of surface water quality monitoring. For chemical analysis, sediments should be digested. Dried sediment samples are digested following the USEPA 3051 guideline or similar methods. About 0.1 g of sediment is digested in 2 mL of HNO_3 (65%) and 0.6 mL of HF (48%) in Teflon bombs using a microwave oven. If the most reactive or bioavailable fraction is to be determined, then cold dilute acid (0.5 to 1.0 M HCl at a ratio of 1:50, sediment to acid) is used for 1 hour. The concentrations of the liquid phase sample are measured by a similar analyzer used for water. In general, the concentrations of trace metals in sediments are higher than those in water. ICP-AES and AAS are used.

The most detailed guideline for sediment quality may be provided by the criteria in Quebec (see Appendix A). It includes the guidelines for fresh water and sea sediment quality. In the criteria, five reference values are provided to protect aquatic

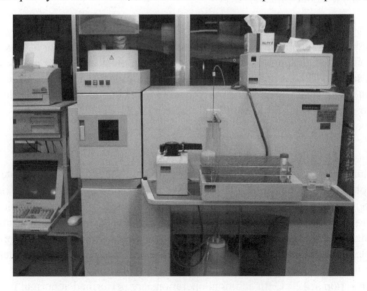

FIGURE 4.10   Photo of an ICP-AES.

life, on eight metals and metalloids (i.e., arsenic, cadmium, chromium, copper, lead, mercury, nickel, and zinc) and twenty-seven organic compounds: a threshold effect level (TEL) and a probable effect level (PEL). These two values have been adopted for the assessment of sediment quality in Quebec, and three other levels were derived to define all of the intervention levels needed for sediment management in Quebec under a variety of contexts. The three new sediment quality criteria were defined using the Canadian Council of Ministers of the Environment (CCME) database and a calculation method similar to the one used to determine the TEL and the PEL. They are (1) the rare effect level (REL), (2) the occasional effect level (OEL), and (3) the frequent effect level (FEL). The criteria can be available for the preservation and protection of sediment quality to protect aquatic life.

Although there are currently no guidelines for speciation of metals, the oxidation states of arsenic and chromium (in particular) can and should be determined because As(III) and Cr(VI) are more toxic and mobile than the As(V) and Cr(III) forms. Arsenic speciation can be determined by hydride generation and atomic spectrometry or by LC-ICP-MS. Chromium speciation can be determined by colorimetric methods or ion exchange methods, coprecipitation, graphite furnace atomic absorption spectrometry (GF-AAS), ICP-AES, or ICP-MS. Other techniques are shown in Figure 4.11.

X-ray absorption spectroscopy (XAS) is an element-specific technique for characterizing electronic configurations at the surface of both amorphous and crystalline materials. It probes the unoccupied electronic structure of a solid, providing structural information similar to X-ray diffraction (XRD) and electronic information

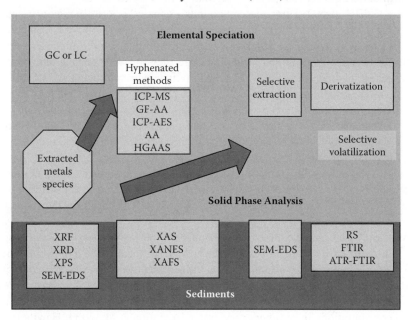

FIGURE 4.11 Summary of techniques used for elemental and metal speciation analysis of sediments.

similar to X-ray photoelectron spectroscopy (XPS). It is essentially a local diffraction approach that can be used to study short-range ordered materials, thus suited for characterizing adsorbed species on poorly crystallized materials. The element specificity of XAS makes it useful to speciate trace elements such as arsenic adsorbed to pure minerals, soil, and sediments (Wilkin and Ford, 2006). Spectra can be collected under in situ or ex situ conditions at ambient pressure.

Two distinct parts of the XAS spectrum are extended X-ray absorption fine structure (EXAFS) and X-ray absorption near edge structure (XANES). EXAFS gives information about the coordination number and interatomic distance, and the nature and position of the neighboring atoms in the coordination shell of the adsorbed ion, which are useful to identify and quantify major mineral phases, adsorption complexes, and crystallinity. XANES spectra yield electronic and structural information with regard to the adsorbed ions and are often used to determine the oxidation state and the local electronic structure within a sample. EXAFS spectra can be collected from materials with arsenic concentrations as low as 100 to 500 mg/kg, whereas that for XANES can be 50 mg/kg or even lower under optimal conditions (Paktunc et al., 2003; Sherman and Randall, 2003).

4.4.3.8 Toxic Substances—Organic Micropollutants

Polychlorinated biphenyls (PCBs) are a mixture of synthetic and organic chemicals and were once widely used in electrical equipment, heat transfer systems, specialized hydraulic systems, and other industrial products. PCBs are highly toxic and potent carcinogens. Therefore, any hazardous wastes that contain more than 50 parts per million of PCBs are subjected to regulation in many countries.

The USEPA lists 116 organic compounds as toxic "priority pollutants;" many states have longer lists. One of the major groupings is volatile organic compounds (VOCs), many of which are chlorine-containing solvents. Lower-molecular-weight VOCs are rarely found in sediments, because they are very soluble and volatile. Chlorinated hydrocarbons such as dichlorodiphenyltrichloroethane (DDT) and 2,3,7,8 tetrachlorodibenzo-p-dioxin (2,3,7,8-TCDD) are very toxic and, due to their persistence, are often found in sediments. There are also petroleum hydrocarbons and starting materials for plastics, dyes, and pharmaceuticals. The "semivolatile" group includes solvents. Pesticides including insecticides, herbicides, rodenticides, and fungicides are highly diverse, persistent, and toxic compounds found in sediments. PAHs (polycyclic aromatic hydrocarbons, like naphthalene and anthracene, which are coal tar constituents) were formerly used in electrical transformers and other products and are potent carcinogens and toxic. They are also the subject of many substance priority lists.

Most of these organic compounds are analyzed routinely by gas chromatography (GC), often followed by mass spectrometry (MS) for identification. Solvent extraction is used to remove the substances from the sediments. Preconcentration by liquid extraction or solid phase microextraction (SPME) may also be necessary. SPME techniques may also be used in situ or ex situ to determine contaminant (PAH and PCB) concentrations in sediment pore water (Azzolina et al., 2009; Hawthorne et al., 2009). Other types of detectors can be used as described by the USEPA. The principal one is photoionization detection (PID) for analysis of VOCs and benzene, toluene, ethylbenzene, and xylenes (BTEX). Flame Ionization detectors (FID) and electron capture detectors (ECD) are applicable for analysis of chlorinated pesticides. GC-mass spectrometric analysis is also possible for detection since quadrupole mass spectrometers are now portable. Coupling with mass spectrometers allows for more specific detection of the compound.

There are also several detectors available which are used alone or in combination with HPLC or other equipment for analysis of the contaminants. The most common are ultraviolet (UV), conductivity, and refractive index (RI). UV detectors are based on the absorption of UV. If a molecule is not UV absorptive, derivatives can be formed to enable UV detection. PAHs can be detected by UV or by fluorescence detectors, which induce fluorescence by a laser light source. Refractive index detectors employ two light beams that are focused on two photocells. One passes through a solvent without contaminants, whereas the other has a different refractive index due to the presence of the contaminant. This type of detector is used for organic acids and many types of saccharides and polysaccharides. Its level of detection is 200 µg/L. Electrochemical detectors detect components at concentrations as low as 1 mg/L. Contaminants are either oxidized or reduced in the cell. Metal ions, in particular, are detected by this type of detector. Ether, carboxylic acids, and alcohols cannot be detected unless they are derivatized. Many components in surface sediments are unknown and may be detrimental to humans if consumed through the food chain. For determination of semipolar and polar contaminants, high performance liquid chromatography–mass spectrometry (HPLC/MS) is the most applicable. Unknowns such as surfactants, pesticides, pharmaceuticals, and herbicides can be determined by LC/MS/MS. Polycyclic aromatic hydrocarbons (PAHs) such as naphthalene and methylnapthalene are common at contaminated sites. They are difficult to detect without derivatization prior to GC/MS, which is more appropriate for semivolatile and volatile compounds. Electrospray ionization (ESI) mass spectrometry techniques have been coupled with liquid chromatography for analysis of PAHs in water and groundwater. A triple quadrupole MS/MS coupled with LC was used to monitor the metabolites of the anaerobic degradation of naphthalene and other PAHs with limits of detection in the range of 200 ng/L for 20-mL samples. Approximately, 1 hour of sample preparation and analysis were required. A recent intensive survey was performed by the U.S. Geological Survey of over 139 streams in 30 states from 1999 to 2000 (Kolpin et al., 2002). Two of the five analytical techniques used involved solid phase extraction (SPE) with single quadrupole LC/MS-ESI. A total of 29 antibiotic compounds were targeted by these methods. Median concentrations were less than 1 µg/L.

The criteria for the assessment of sediment quality in Quebec (see Appendix A) provide the concentrations for FEL: frequent effect level, OEL: occasional effect level, PEL: probable effect level, REL: rare effect level, and TEL: threshold effect level, for 27 organic compounds, besides metals and metalloids. They are total polychlorinated biphenyls (PCBs), nonylphenol and its ethoxylates, PCDD/PCDF (ng tox eq/kg), acenaphthene, acenaphthylene, anthracene, benzo[a]anthracene, benzo[a]pyrene, chrysene, dibenzo[a,h]anthracene, fluoranthene, fluorine, 2-methylnaphthalene, naphthalene, phenanthrene, pyrene, chlordane, DDD, DDE, DDT, dieldrin, endrin, heptachlor, epoxide, lindane, and toxaphene. The criteria can be available for the preservation and protection of sediment quality. Protocols for sampling and storage should be verified with an analytical laboratory.

4.4.3.9 Other Environmental Indicators

Radioactivity (e.g., total alpha and beta activity, ^{137}Cs, ^{90}Sr), microbiological (e.g., total coliforms, fecal coliforms, viruses, yeasts, parasites, and fecal streptococci bacteria), and biological indicators (e.g., phytoplankton, zooplankton, zoobenthos, fish, macrophytes, and birds and animals related to surface waters) are often included as the objects of surface water quality monitoring. Microbial indicators in the sediments are particularly important near sewage discharges and wastewater outfalls. They are of concern near beaches, shellfish beds, and drinking water intakes.

Due to the deficiencies in the standard methods for detection of viruses, bacteria, and protozoa, particularly the length of time, various methods have been developed including immunofluorescent antibodies techniques, fluorescent in situ hybridization, magnetic bead cell sorting, electrochemiluminescence, amperometric sensors, and various polymerase chain reactions (PCR), RT-PCR, and real-time PCR methods. For example, Hoostal et al. (2002) performed PCR and RT-PCR on DNA and RNA from sediments of the western basin of Lake Erie of the Great Lakes. Magnification of the bphA1 gene could be useful as a screen for catabolic activities for PCB mixtures in sediments.

Biological monitoring involves the determination of the numbers, health, and presence of various species of algae, fish, plants, benthic macroinvertebrates, insects, or other organisms as a way of determining water quality and bioavailability of the contaminant. Bioavailability as defined by the NRC (2003) is "the individual physical, chemical and biological interactions that determine the exposure of plants and animals to chemicals associated with soils and sediments." Knowledge of background information is essential. Attached algae (known as periphyton) are good indicators of water quality because they grow on rocks and other plants in the water. Advantages are that high numbers of species are available, their responses to changes in the environment are well known, they respond quickly to exposures and are easy to sample. An assessment could include determination of the biomass by chlorophyll or on an ash-free dry basis, species, distribution of species, and condition of the attached algae assemblages. Their use has not been widely incorporated yet in monitoring programs.

Benthic macroinvertebrates have numerous advantages as bioindicators (protocols). They do not move very far and thus can be used for upstream–downstream studies. Their life span is about a year, enabling their use for short-term environmental changes. Sampling is easy, they are numerous, and experienced biologists can easily detect changes in macroinvertebrate assemblages. In addition, different species respond differently to various pollutants. They are also food sources for fish and other commercial species. Many states in the United States have more information on the relationship between invertebrates and pollutants than for fish.

Aquatic plants (macrophytes) grow near or in water. A lack of macrophytes can indicate quality problems caused by turbidity, excessive salinity, or the presence of herbicides. Excessive numbers of these plants can be caused by high nutrient levels. They are thus good indicators, because they respond to light, turbidity, and contaminants such as metals, herbicides, and salt. No laboratory analysis is required, and sampling can be done through aerial photography.

Biosurveys are useful in identifying if a problem exists. Chemical and toxicity tests would then be required to determine the exact cause and source. Routine biomonitoring can be less expensive than chemical tests over the short term but more expensive over the long term. Field bioassessment experts are required to obtain and interpret data, and there are no established protocols. More knowledge is required to determine the effects of contaminants on populations of organisms and better coordination of background data before site contamination.

Acute and chronic toxicity bioassays, bioaccumulation bioassays, and biomarkers tests can all be performed to determine the toxicity of the contaminants in the environment. Biomarkers such as P450 are used. In situ or lab short-term toxicity bioassays examine the median lethal concentration (LC50) that would kill 50% of the organics. Tests are performed with specific species over the period of hours to days. Chronic tests are used to examine long-term effects (over several weeks) which are physiological, pathological, immunological, teratological, mutagenic, and carcinogenic. The assays are performed on water column organisms that are affected by suspended solids or benthic organisms that live in the deposited sediments. The suspended solids or sediments are homogenized and placed in replicate containers. Reference and controls are also designated. The organisms are counted and added. After 10 days, the containers are emptied, and the organisms are counted. The results are statistically compared to the reference materials to obtain the toxicity effect. Various biases must be addressed, such as the sensitivity of some organisms to fine sediments or high TOC content in the sediments. Short-term toxicity tests are often criticized, because field exposures are mainly chronic. The tests can also be expensive and difficult to reproduce and perform.

Various species may also be used to evaluate bioavailability of contaminants and potential accumulation in the food web. Tests are performed from 10 to 28 days by measuring the contaminants in the tissues of the organisms. Species selection is key. The USEPA/USACE (1991, 1998) and PIANC (2006) have provided guidelines for evaluating the toxicity. Two or three species are usually used.

Acid volatile sulfide–simultaneously extracted metals (AVS:SEM) models as previously described are also used to simulate bioaccumulation for metals. For organic molecules, k_{ow} and sediment organic content data can be used for screening purposes.

4.4.3.10   Test Kits

Test kits are available from various manufacturers for the determination of a wide variety of analytes in surface water. ASTM D 5463-93 lists some types that are available for 44 inorganic analytes. The range, detection limits, sensitivity, accuracy, and susceptibility to interferences vary from kit to kit, depending on the methodology selected by the manufacturers (i.e., appearance/turbidity, visual colorimetric, go/no go, photometric, and titrimetric).

Reagent kits are designed for analysis of water. The equipment required for each kit can vary widely from a color wheel or chart to an electronic spectrophotometer to detect color change. Most kits include containers, reagents, and calibration standards. Reagent kits can be used for individual or general classes of compounds. Some contaminants detected by the kits include polychlorinated biphenyls (PCBs), pol-yaromatic hydrocarbons (PAHs), various oils and fuels, benzene, toluene, ethylben-zene, and xylenes (BTEX), trihalomethanes, and explosives such as trinitrotoluene (TNT), cyclotrimethylenetrinitramine (RDX), and cyclotetramethylenetetranit-ramine (HMX).

4.5 DECISION MAKING USING INDICATORS

Investigation of sediment quality can be separated into three cases: regular monitor-ing on sediment quality, eutrophication, and contamination with toxic or hazardous substance(s). Regular monitoring may primarily consist of the selections of items to be monitored and sites and frequency for monitoring. Selection of sampling methods is also important. The evaluation can be made from the results of analyses. Reports should describe the changes in the environmental indices, as shown in Figure 4.12. If the level is over the criteria, the measures can be considered.

There are many sites where eutrophication is serious, where both the sur-face water and sediment quality are concerned. In general, eutrophication, SS, COD, T-N, T-P, chlorophyll a, etc. can be measured for surface water quality. For sediments, some physical properties such as redox potential can be measured, as shown in Figure 4.13. If sediment is rich in nutrients and releases the nutrients, measures for control can be considered. In many cases, dredging or capping has

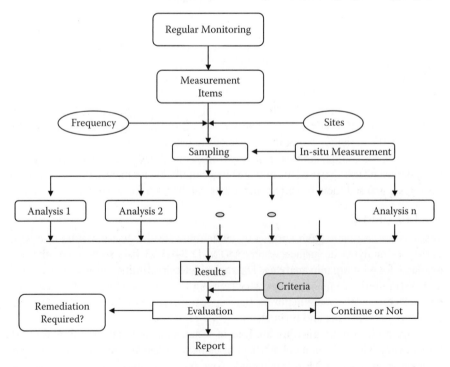

FIGURE 4.12 Decision-making process for regular monitoring of sediment quality.

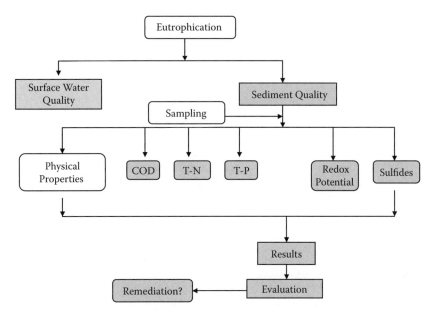

FIGURE 4.13   Decision-making process for the investigation of sediment eutrophication.

been performed, and the results have been evaluated as in some cases described in Chapter 7.

If sediments are contaminated with toxic or hazardous substance(s), the level of the targets has to be appropriately measured. The results from analyses and the criteria are used for the evaluation. The most important decision is if remediation is required or not, as shown in Figure 4.14. For the evaluation, risk analysis may be feasible.

To predict pore water concentration, equilibrium partitioning (EqP) models as seen previously have been used.

$$C_p = \frac{c_s}{f_{oc}K_{oc}} \tag{4.9}$$

where C_p is the pore water concentration in µg/L, c_s is the sediment concentration in µg/kg, f_{oc} is the fraction of organic carbon in the sediment (=%TOC/100), and K_{oc} is the organic carbon/water partition coefficient. K_{oc} can be estimated from the Karickhoff et al. (1979) equation presented previously (Equation 3.7).

4.6   CASE STUDIES

4.6.1   Investigation of Port Sediments

Shimizu Port is one of the 23 specifically designated important ports in Japan. Orido Bay is located in the inner part of Shimizu Port. The bay has been used as a pool in which wood has been floated for pest control since 1927. However, since the bay became surrounded by many industries, it might have become contaminated. An

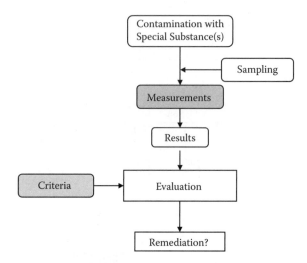

FIGURE 4.14 Decision-making process for the investigation of sediment contamination.

integrated investigation was, therefore, needed to assess its potential for the following uses: the farming of fish and shellfish, tourism, anchorage, moorage and roadstead areas for ships and seaplanes, or a waterfront public park.

The investigation sites were within a wood pool that occupies most of Orido Bay and Shimizu Port. The wood pool has an area of 717,000 m² and a maximum water depth of 8 m. Therefore, wood chips have accumulated on the bottom and have been subjected to degradation. The pool is fed by two streams, the Ohashi River and the Hamada River, both of which run through residential districts and small factory sites, as shown in Figure 4.15. The flow rates of the streams are relatively low (i.e., 0.17 to 0.76 m³/s and 0.02 to 0.24 m³/s, respectively). The Tomoe River, which is larger than the two streams mentioned above, flows into Shimizu Port. The flow rate of the Tomoe River was approximately 6.0 m³/s. The properties of discharged suspended solids and sediments were also determined for the samples from each of the river mouths and are shown in Table 4.2 (Sato et al., 2006).

Nineteen surface sediment samples were obtained using a grab sampler from the bottom of the wood pool (Orido Bay) and Shimizu Port (Figure 4.15). The sampling depth was less than 8 cm from the top surface. As an example, the data of ignition loss for Orido Bay sediments are plotted for the same site in Figure 4.16. It shows the relationships between ignition loss and distance from S1 site, located just outside of Orido Bay. The distance between S1 and S7 is approximately 5000 m. As can be seen in Figure 4.16, the ignition loss of the sediments decreases with distance from the S1 site. This indicates that the source of the organic matter is mainly within Orido Bay. The ignition loss of soils can represent the organic matter content.

Because organic matter can contain more water, a higher ignition loss usually leads to higher water content. Furthermore, a lower specific gravity of organic matter can also make the water content higher. A similar pattern is seen in peat, which

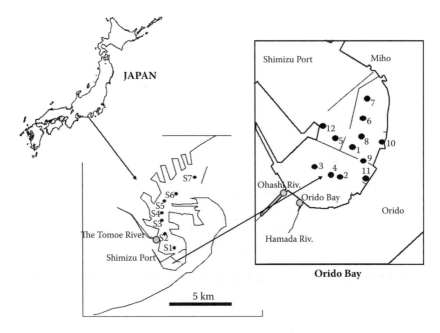

FIGURE 4.15　Sampling sites for Shimizu Port, Japan.

consists mainly of organic matter. The relationship between ignition loss and water content is shown in Figure 4.17.

As a low *Eh* value (reduction potential) means a reducing anaerobic condition, the sediments of S1, S2, and Orido Bay are now under anaerobic conditions. Figure 4.18 shows that sediments with a low *Eh* value contain a high concentration of sulfide. The sulfide concentration was very low at S4 to S7, where *Eh* values were relatively high.

TABLE 4.2

Comparison of Heavy Metal Concentrations for SS and Sediments

Sites	Water Flow (m³/s)	LOI (%)	Metal Content (mg/kg dry basis)			
			Cu	Pb	Zn	Cd
Hamada R.	0.02–0.24					
SS		2.5	177	102	480	0.36
Sediments		7.0	164	42	318	0.34
Ohashi R.	0.17–0.76					
SS		6.9	255	301	952	2.35
Sediments		3.1	52	27	169	0.04
Tomoe R.	6.0					
SS		2.0	150	27	214	0.13
Sediments		9.6	79	27	239	0.35

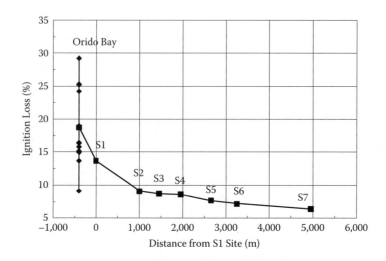

FIGURE 4.16 Ignition loss in port sediments. Vertical line indicates samples from Orido Bay.

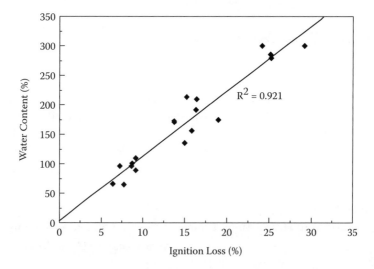

FIGURE 4.17 Water content and ignition loss for surface sediments in Shimizu Port.

The organic component of sediments has a high affinity for heavy metal cations because of the presence of ligands or groups that can form chelates with the metals (Yong et al., 1992). If this is the case, the heavy metal contents must increase with increasing ignition loss. Figure 4.19 shows the concentrations of copper, lead, and zinc versus the ignition loss for sediment samples from Orido Bay (Fukue et al., 2006a) and Shimizu Port (Fukue et al., 2007). Thus, the metal concentrations increased with increasing ignition loss (IL). Because the background value of Zn is 130 mg/kg dry for silty sediments (Fukue et al., 2006b), the rate of increase in the

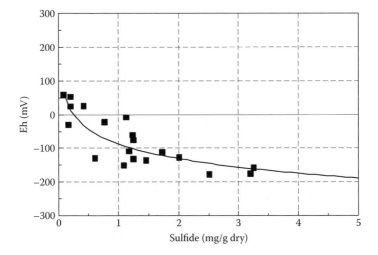

FIGURE 4.18   Correlation between *Eh* and sulfide for surface sediments from Shimizu Port.

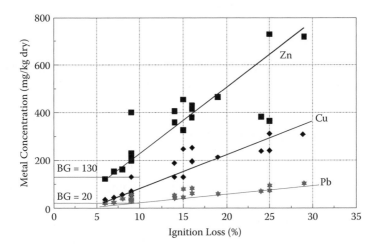

FIGURE 4.19   Relationships between metal concentration and ignition loss. BG = background.

amount of Zn due to ignition loss is approximately 28 mg/kg/%. The background value for both Cu and Pb is 20 mg/kg (Fukue et al., 2006b).

4.6.2   LAKE SEDIMENTS

Lake Sanaru is located in the western part of the city of Hamamatsu, Shizuoka Prefecture. The area of the lake is 1.2 km², and the basin has an area of 17.3 km², where the population has increased rapidly for the last 30 years. For example, in

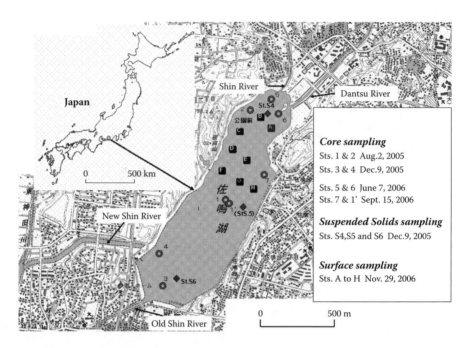

FIGURE 4.20   Sampling sites of sediments in a brackish lake in Japan.

1965, the basin was wilderness. By 1993, 66% of the area had become residential land. The wastewater from the residences contained high levels of COD and total nitrogen (T-N).

This is considered to be a main reason for the decreasing quality of surface water in the lake. As a result, since 2001, the water quality in Lake Sanaru is the worst in Japan. A further understanding of the contamination mechanisms and technical requirements for remediation are desired by the local government and citizens.

Core samples were obtained by penetrating a stainless thin-walled tube with a length of 1 m and a diameter of 86 mm at several site locations, as shown in Figure 4.20. The lengths of the samples obtained ranged from 40 to 75 cm. The samples were maintained upright for transportation and storage. The core samples were sliced in intervals of 1, 3, or 25 cm, depending on the objective of the investigation. For dating, ^{137}Cs and ^{210}Pb methods were used on samples with a thickness of 1 cm.

The concentrations of ^{137}Cs were measured on the core sample obtained from St. 1. The result is shown in Figure 4.21. The figure shows that the ^{137}Cs peak appears at a depth of around 20 and 22 cm. The peak is due to the fallout of the atmospheric bomb test around 1960 (Callaway et al., 1996). Thus it is deduced that sediment age at a depth of 22 cm is 43 years (1963–2006).

The use of ^{210}Pb dating is increasing rapidly (Appleby et al., 1979). In this study, the ^{210}Pb dating technique was also used. The result in Figure 4.21 shows that the concentration of ^{210}Pb decreased almost linearly with mass depth in relatively shallow layers. It is noted that a mass depth of 10 g/cm^2 is about 25 cm in depth. The

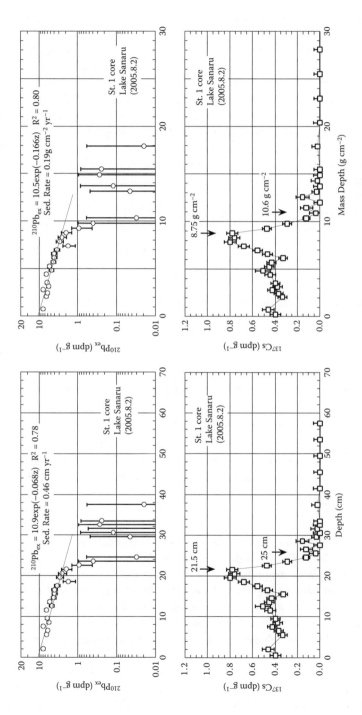

FIGURE 4.21 ^{137}Cs dating on a core sample from St. 1 (at the center of the lake) and determination of the sedimentation rate using the ^{210}Pb method.

linear decrease of the ^{210}Pb concentration may indicate that the upper layer was not disturbed, while lower sediments have been disturbed for some reason.

Since ^{210}Pb concentration decreases with the age of sediments, a large scattering of data beyond a mass depth of 9 g/cm^2 can be due to the disturbance of the sediments. The disturbed concentrations of ^{210}Pb plotted in Figure 4.21 may be too low in comparison to the extension of the straight line obtained, which may indicate that a layer of the sediments was removed by natural or artificial events before the upper sediments has been deposited. This is discussed later. From the ^{210}Pb dating technique, sedimentation rate was estimated to be 0.18 g/cm^2/y.

Using the dating from ^{137}Cs and ^{210}Pb data, the sediment ages were estimated. The average particle density was used for the total sediment layer. Basically, the time (in years) required for sedimentation of 1 cm sediment layer was calculated using the following equations:

The definition of water content w is

$$w = \frac{w_w}{w_s} \times 100\% \tag{4.10}$$

where w_w and w_s are the mass of water and solids, respectively. Void ratio e is defined as

$$e = \frac{V_v}{V_s} \qquad V_s + V_v = 1 \tag{4.11}$$

where V_v and V_s are the volume of void and solid, respectively.

$$e = wGs / 100 \qquad w_s = V_s \rho_s \tag{4.12}$$

where G_s is the specific gravity of solid, and ρ_s is the density of solid. After w_s is obtained, the time required for the sedimentation of a 1-cm-thick layer is calculated by

$$\text{time} = w_s/\text{sediment.rate} \tag{4.13}$$

The calculation shows where the bottom surface is estimated to be for the sediments in 2007. Thus, it is deduced that dredging was not performed around St. 1.

It was shown, from the ^{210}Pb results, that there are disturbed layers which possibly exist below a depth of 26 cm. This depth was estimated to be around 1950. There were two historical events in those days. One is the occurrence of the M 8.0 earthquake (Tou Nankai Earthquake in 1944), and the other was the air strikes by the U.S. Air Force. These events would potentially disturb the surface sediments.

Figure 4.22 through Figure 4.25 show the profiles of zinc, copper, lead, and cadmium concentrations, respectively, for the core samples obtained from the middle part of Lake Sanaru. The background values seem to be around 100, 20 to 30, around 20, and 0.5 to 0.2 mg/kg for Zn, Cu, Pb, and Cd, respectively, as indicated in the figures, where the background value is defined as the concentration level without

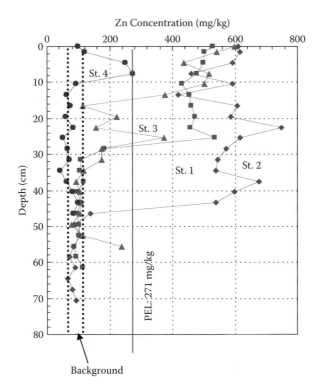

FIGURE 4.22   Zn concentration profile in brackish lake sediments.

anthropogenic effects. It was found that the surface layers of the sediments were considerably contaminated with these heavy metals, compared to the background values. The frequent effect level (FEL) for Zn is 770 mg/kg for freshwater and 430 mg/kg for marine sediments (Environment Canada and Ministère du développement durable, de l'environnement et des Parcs du Québec (2008)). As the sediments are brackish, either criterion may be used. If the guideline for marine sediments is applied, the Zn concentration exceeds the FEL.

At St. 1, the Zn concentration abruptly changes at depths ranging from 30 to 26 cm, as shown in Figure 4.22. The dating analyses using the ^{137}Cs and ^{210}Pb showed that the sediment was deposited around 1945. It is likely that the Zn concentration started to increase around 1950, which accords with time for the beginning of a population increase near Lake Sanaru. The present population is approximately seven times that of 1950.

A relatively low concentration of heavy metals at St. 4 may be due to dredging effects. The contaminated sediments would have been removed in 1992. The top surface, with a thickness of a few centimeters, was replaced by sand. Sediments having a relatively high concentration at a depth of 8 cm may have been deposited when pollution started around 1950. There is a correlation between variables, and the intensity of correlation depends on the variables. Some correlations between items are presented in Table 4.3.

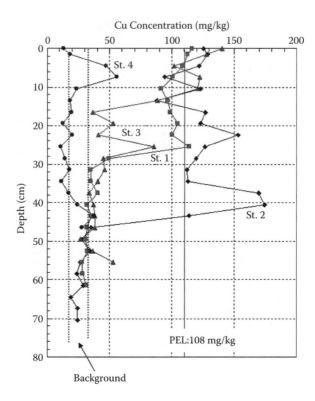

FIGURE 4.23 Cu concentration profile for brackish lake sediments.

With eutrophication, COD and T-N of the surface sediment samples obtained from Lake Sanaru were measured. The values of COD in sediments varied between approximately 10 and 70 mg/g, as shown in Figure 4.26. On the other hand, total phosphorus in sediments ranged between 0.15 and 0.8 mg/g, as shown in Figure 4.27. Thus, COD and T-N can be released into water when organic matter content is degraded.

The study showed that the sediments in the lake contain excess amounts of nutrients, COD, heavy metals, and so on. In order to purify the surface water, the remediation of sediments is inevitably required.

4.7 CONCLUDING REMARKS

The investigation of sediment quality aims at the preservation and protection of environments for aquatic life. Sediment quality for regular monitoring, eutrophication, or contamination can be evaluated by performing physical and chemical analyses, which are usually integrated into the investigation.

The selection of variables to be measured is the most basic planning. The variables may be directly or indirectly related to the monitoring objectives in order to evaluate the results properly. The integrated investigation may spread over many fields, such as geotechnical engineering, geochemistry, chemical analyses (analytical chemistry), biology, microbiology, etc.

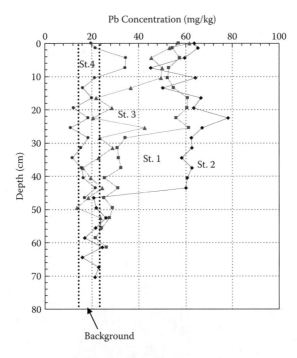

FIGURE 4.24 Pb concentration profile for brackish lake sediments.

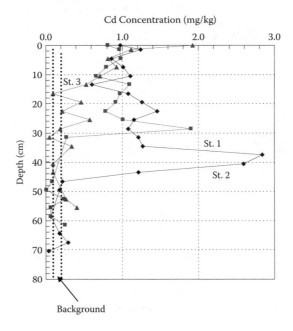

FIGURE 4.25 Cd concentration profile for brackish lake sediments.

TABLE 4.3
Correlation Factors between Index Variables

	Zn	Cu	Pb	Cd	T-S
Zn		0.985	0.864	0.687	0.353
Cu			0.834	0.725	0.412
Pb				0.627	0.127
Cd					0.155
T-S					

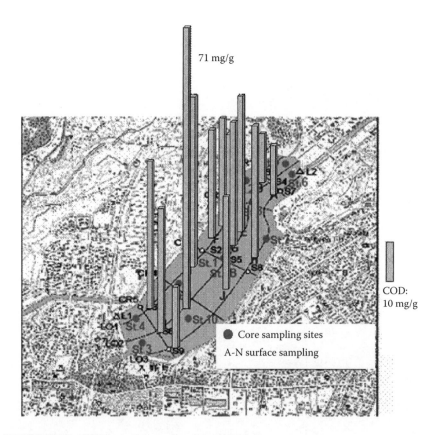

FIGURE 4.26 COD in sediments from a brackish lake.

For sediments, most cases require sampling procedures. Therefore, investigation requires the use of boat, sampler, and divers, and the cost for physical and chemical analyses. The investigation plan should be carefully prepared. For example, weather is also an important factor for the performance of investigation. The plan should not be too tight.

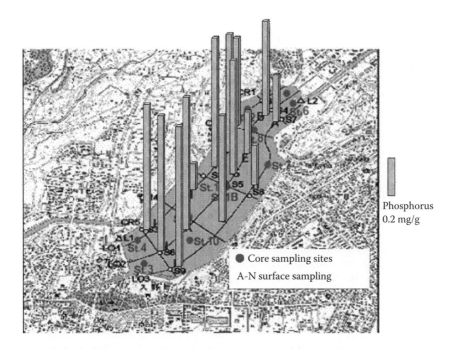

FIGURE 4.27   Total phosphorus in sediments from a brackish lake.

In many cases, external laboratories may be needed for analysis. As the analyses often include errors, accuracy should be checked in an appropriate manner, such as cross-checking.

REFERENCES

Appleby, P.G., Oldfield, F., Thompson, P., Huttunen, P., and Tolonen, K. 1979. ^{210}Pb dating of annually laminated lake sediments from Finland, letters. *Nature* 280: 53–55.

ASTM. 2006. *ASTM Standards on Environmental Sampling*, 3rd version. ASTM International, West Conshohocken, PA.

Azzolina, N.A., Neuhauser, E.F., and Nakles, D.F. 2009. *SPME Pore Water for Characterizing PAH-Impacted Sediments.* 5th International Conference on Remediation of Contaminated Sediments, Feb. 2–5, 2009, Jacksonville, FL.

Bright, D.A., Dushenko, W.T., Grundy, S.L., and Reimer, K.J. 1995. Effects of local and distant contaminant sources: polychlorinated biphenyls and other organochlorines in bottom-dwelling animals from an Arctic estuary. *Sci. Total Environ.* 160/161: 265–283.

Callaway, J.C., Delaune, R.D., and Patrick, Jr., W.H. 1996. Chernobyl ^{137}Cs used to determine sediments accretion rates at selected northern European coastal wetlands. *Limnol. Ocean.* 41(3): 444–450.

Canadian Sediment Quality Guidelines (CSeQGs) 2003. Summary of Existing Canadian Environmental Quality Guidelines, SUMMARY TABLE. http://www.ec.gc.ca/CEQG-RCQE/English/Ceqg/Sediment/default.cfm.

Environment Canada. 2002a. *Sediment Sampling Guide for Dredging and Marine Engineering Projects in the St. Lawrence River. Volume 1, Planning Guidelines.* Environment Canada, Environmental Protection Branch, Quebec Region Technological Innovation and Industrial Sectors Section. Report 101 p.

Environment Canada. 2002b. *Sediment Sampling Guide for Dredging and Marine Engineering Projects in the St. Lawrence River. Volume 2, Field Operations Manual.* Environment Canada, Environmental Protection Branch, Quebec Region Technological Innovation and Industrial Sectors Section. Report 103 p.

Environment Canada and Ministère du Développement durable, de l'Environnement et des Parcs du Québec. 2008. Criteria for the Assessment of Sediment Quality in Quebec and Application Frameworks: Prevention, Dredging and Remediation.

Fukue, M., Mulligan, C. N., Sato, Y., and Fujikawa,T. 2007. Effect of organic suspended solids and their sedimentation on the surrounding sea area. *Environ. Poll.* 149(1): 70–78.

Fukue, M., Nakamura, T., and Kato, Y. 1999. Cementation of soils due to calcium carbonate. *Soil Found.* 39: 55–64.

Fukue, M., Sato, Y., Uehara, K., Kato, Y., and Furukawa, Y. 2006a. Contamination of sediments and proposed containment technique in a wood pool in Shimizu, Japan. *ASTM STP 1482*: Evaluation and Remediation Techniques, ASTM STP1482, 32–43.

Fukue, M., Yanai, M., Sato, Y., Fujikawa, T., Furukawa, Y., and Tani, S. 2006b. Background values for evaluation of heavy metal contamination in sediments, *J. Haz. Mat.,* 136: 111–119.

Furukawa, Y., Fujita, T., Kunihiro, T., and Fukazawa, M. 2001. Investigation of particle size distribution of soil using a particle size analysis equipment automated by laser and its applicability to soil samples (text in Japanese). *J. Geotech. Eng.*, Japan Society of Civil Engineers, 56 (687): 219–231.

Hawthorne, S.B., Miller, D.J., Grabanski, C.B., Ghosh, U., and Kwon, S. 2009. Comparison of in situ POM and ex situ SPME methods to determine PCB concentration in sediment pore water. *5th International Conference on Remediation of Contaminated Sediments,* Feb. 2–5, 2009, Jacksonville, Florida.

Hoostal, M.J., Bullerjahn, G.S. and Mckay, R. M.L. 2002. Molecular assessment of the potential for in situ bioremediation of PCBs. *Hydrobiologia* 469: 59–66.

The Japan Fisheries Resources Conservation Association 2005. Water standards for fishery. http://ay.fish-jfrca.jp/kiban/kankyou/hourei/yousui/suisan_kijyun.html.

Karickhoff, S.W., Brown, D.S., and Scott, T.A. 1979. Sorption of hydrophobic pollutants on natural sediments. *Water Resources*, 13: 241–248.

Kolpin, D.W., Furlong, E.T., Meyer, M.T., Thurman, E.M., Zaugg, S.D., Barber, L.B., and Buxton, H.T. 2002. Pharmaceuticals, hormones, and other organic wastewater contaminants in U.S. streams, 1999–2000: A national reconnaissance. *Environ. Sci. Technol.* 36 (6): 1202–1211.

McLachlan, M.C., Haynes, D., and Müller, J.F. 2001. PCDDs in the water/sediment-seagrass-dugong (*Dugong dugon*) food chain on the Great Barrier Reef (Australia). *Environ. Poll.*113: 129–134.

NRC. 2003. *Bioavailability of Contaminants in Soils and Sediments: Processes, Tools and Applications.* National Research Council, National Academies Press, Washington, DC.

Paktunc, D., Foster, A., and Laflamme, G. 2003. Speciation and characterization of arsenic in Ketza river mine tailings using X-ray absorption spectroscopy. *Environ. Sci. Technol.* 37: 2067–2074.

PIANC. 2006. Biological assessment guidance for dredged material. Report from Working Group 8 of the Environment Commission. International Navigation Association. Brussels, Belgium,

Sato, Y., Fukue, M., Yasuda, K., Kita, K., Sawamoto, S., and Miyata, Y. 2006. Transport and contamination of suspended solids in Shimizu Port. ASTM STP1482, 19–31.

Sherman, D.M. and Randall, S.R. 2003. Surface complexation of arsenic(V) to iron(III) (hydr) oxides: Structural mechanism from abinito molecular geometries and EXAFS spectroscopy. *Geochim. Cosmochim. Acta* 67: 4223–4230.

Thompson, M., White, T., and Graves, G. 2001. Degradability of sediments from the St. Lucie Estuary Florida: A pilot study. Florida Department of Environmental Protection, Water Quality Program, Port St. Lucie. http://www.dep.state.fl.us/southeast/ecosum/ecosums/ Degradability_Sediments_SLE.pdf.

USEPA/USACE. 1991. Evaluation of dredged material proposed for ocean disposal. *Testing Manual: Inland Testing Manual.* U.S. Environmental Protection Agency/U.S. Army Corps of Engineers EPA-503-B-91-001. U.S. EPA Office of Water (556F), Washington, DC.

USEPA/USACE. 1998. Evaluation of dredged material proposed for discharge in waters of the U.S. *Testing Manual: Inland Testing Manual.* U.S. Environmental Protection Agency/ U.S. Army Corps of Engineers EPA-823-B-98-004. U.S. EPA Office of Water (4305), Washington, DC.

Wilkin, R.T. and Ford, R.G. 2006. Arsenic solid-phase partitioning in reducing sediments of a contaminated wetland. *Chem. Geol.* 228: 156–174.

Yong, R.N., Mohamed, A.M.O., and Warkentin, B.P. 1992. *Principles of Contaminant Transport in Soils.* Elsevier, Amsterdam.

5 Natural Recovery of Contaminated Sediments

5.1 INTRODUCTION

Natural attenuation involves the use of the natural processes with the soil and groundwater to remediate contamination by physical, chemical, and biological processes to reduce the risk to human health and the environment. Although the use of natural attenuation as a treatment process is increasing for remediation of contaminated groundwater, much less research has focused on contaminated soils and sediments. Industrial effluents, agricultural runoff, and sewage discharges are major sources of contaminants for the sediments. In addition, benthic organisms can transport contaminants through bioturbation, and there is considerable variability at sites. Organic matter, a particularly important component of the sediments, can sequester the contaminants. Sediment–water partitioning controls the release of the contaminants into pore water and benthic organisms. Fate and transport mechanisms for both organic and inorganic contaminants within the sediments need to be understood to establish protocols for the monitoring and use of natural attenuation.

A summary of the fate and transport processes for various environmental scenarios is presented in Figure 5.1. The parameters including pH, oxidation–reduction conditions, amount of organic matter, salt content, mixing energy, and potential interaction with groundwater vary according to the environment. These parameters subsequently influence the type of dominant fate and transport processes.

The contaminants that are left in the sediment environment without intervention can undergo naturally occurring processes. It is often more cost-effective than dredging, capping, or treatment or combinations thereof, of the contaminated sediments and can be appropriate for low-risk areas. In 1994, the United States Environmental Protection Agency (USEPA) included natural recovery as part of the Sediment Management Strategy (USEPA, 1994). It is "not acceptable where contamination poses severe and substantial risks to aquatic life, wildlife and human health. It may not be the method of choice for contaminants that biodegrade or transform into more persistent, toxic compounds."

Source control is key in preventing recontamination and ensuring the sustainability of the remediation. Lead was a major contaminant due to leaded gasoline use until its ban. Since 1973, annual lead emissions in the United States decreased from 200,000 to 500 tonnes (Callendar, 2005). Reservoir and lake sediment cores have shown decreasing lead levels since 1975. This is in contrast to other metals such as Cu, which is still increasingly released due to metal production. This is reflected in sediment cores, which have shown increased Cu trends.

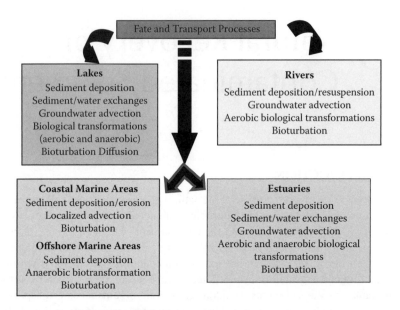

FIGURE 5.1   Fate and transport of contaminants in various environments.

Like natural attenuation of soil and groundwater, monitoring is a key element in determining the success and applicability of this remediation method for mitigating contaminants in the sediments. The natural processes include biological degradation, volatilization, dispersion, bioturbation, dilution, radioactive decay, sorption of the contaminant onto the organic matter and clay minerals in the sediments, and natural capping processes (Figure 5.2). These mechanisms will be discussed further later on.

Natural attenuation is mainly used for benzene, toluene, ethylbenzene, and xylene (BTEX) and more recently for chlorinated hydrocarbons. Other contaminants that could potentially be remediated by natural attenuation include pesticides, polychlorinated biphenyls (PCBs), and inorganic compounds (Yong and Mulligan, 2004). The success of natural attenuation depends on the site conditions, sediment characteristics, potential for downstream transport, and microbiology.

In this chapter, the mechanisms involved and case studies of natural recovery of various pollutants at contaminated sediment sites will be examined. There are differences in the type of processes that play a role in the natural attenuation of groundwater and the natural recovery of sediments. Usually transformation processes are more dominant in natural attenuation, whereas isolation and mixing are more prevalent for sediments. The term "natural recovery" was defined by the Oregon Department of Environmental Quality (DEQ, 1998). It includes both attenuation aspects (reduction of contaminants with no transport to other media) and recovery (which allows the benthic and pelagic communities to be reestablished and resume their beneficial uses). Monitoring is required to ensure that the remediation objectives are achieved and that it is proceeding as planned. Thus the term MNR, monitored natural recovery, is used. Thus upon successful completion, MNR would meet the needs of sustainability.

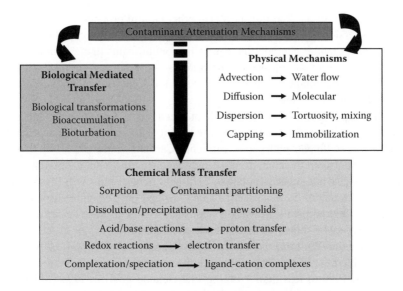

FIGURE 5.2 Mechanisms of natural recovery in sediments.

According to the NRC (2007), MNR is defined as "leaving contaminated sediments in place" and allowing the ongoing aquatic, sedimentary, and biological processes to reduce the bioavailability of the contaminants in order to "protect receptors." The USEPA (2005) definition is slightly different. It says that MNR employs the ongoing, naturally occurring processes to contain, destroy, or otherwise "reduce the bioavailability or toxicity of contaminants" in sediment. Monitoring is included to assess that the risk is reduced as expected.

Currently, U.S. EPA views MNR as a mature technology (USEPA, 2005). More than 28 projects in the United States recently took place where MNR was the primary technique used or it was used in combination with other technologies. It sometimes, however, is difficult to persuade the regulatory agencies and the public to use MNR because the contaminants are left in place. Acceptance is increasing because there are substantial costs associated with the removal of large volumes of sediments. Risks to humans and the environment can be substantial (Wenning et al., 2007). There are still many gaps in knowledge, and a careful evaluation of the options must be made.

5.2 NATURAL RECOVERY PROCESSES OF SEDIMENTS

According to the NRC (1997), MNR involves the allowance of the aquatic, sedimentary, and biological processes to reduce the bioavailability of contaminants to protect receptors. These mechanisms have been previously discussed in Chapter 3. Preference for natural recovery is given to transformation processes that convert contaminants to less toxic forms. Sorption and binding processes enable the reduction of contaminant mobility and bioavailability. Formation of metal sulfides under anaerobic conditions reduces the solubility of the metals and thus the risk to organisms.

Contaminant release and transport is another natural process. This can be due to particle dispersion and diffusive/advective transport. Bioturbation and gas ebullition (due to organic matter degradation) can bring contaminants to the sediment surface for potential subsequent release. Bioturbation is a more important mechanism with an effective contaminant release coefficient (k_{eff}) of 1 cm/yr compared to gas ebullition (0.0033 cm/yr) (Reible, 2006).

Natural capping is one of the dominant mechanisms in sediments and enables reduction of the exposure of the surface sediment. This involves the covering of the contaminated sediments with clean sediments (Cardenas and Lick, 1996), thus forming a barrier between the contaminated sediments and the aquatic environment and reducing toxicity and bioaccumulation. Sediment deposit rates will thus determine the rate of attenuation in this case, and they depend on erosion, resuspension, and the source of the sediments. The newly deposited sediments must be clean and, upon deposit, must not be subject to resuspension and erosion.

At the Sagamo-Weston/Lake Hartwell/Twelvemile Creek Superfund Site natural capping enabled the PCB contaminants on the surface sediment to be reduced from 40 mg/kg in 1979 to less than 1 mg/kg (the cleanup goal) in 1999 (Magar et al., 2004). River sedimentation is highly variable due to seasonal variations in the flow of the river. The rate of sedimentation and initial level of contamination will determine if the quality standards can be met in an acceptable time frame. Severe floods or ice movement can erode the newly deposited sediments.

As sedimenting particles adsorb contaminants, they need to be characterized and monitored (Figure 5.3). The sources of the particles can be urban overflows, storm drains, or runoff from agricultural and thus can be highly variable and hard to control. Contaminants within the sediment can slow the remediation process.

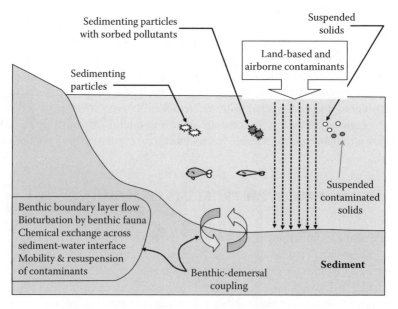

FIGURE 5.3 Settling of suspended solids and other physical transport mechanisms.

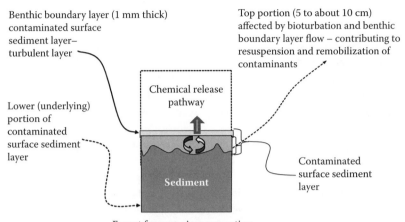

Benthic boundary layer (1 mm thick) contaminated surface sediment layer– turbulent layer

Top portion (5 to about 10 cm) affected by bioturbation and benthic boundary layer flow – contributing to resuspension and remobilization of contaminants

Chemical release pathway

Lower (underlying) portion of contaminated surface sediment layer

Sediment

Contaminated surface sediment layer

Except for excessive wave action, contaminants in lower portion of contaminated sediment layer will likely remain physically undisturbed

FIGURE 5.4 Bioturbation and other physical processes in the benthic layer.

Propellers, waves, and the benthic community may all cause the disturbance of sediments. Clean sediments that have covered contaminated sediments may be removed, thus reexposing the contamination. The benthic community can also bring the contaminated sediments to the surface through working the sediments or enhancing the pore water exchange rate. The benthic activity may cause the oxidation of the sediments. The oxidation of the sediments may release heavy metals bound to sulfides in previously anaerobic sediments (USEPA, 1995). The mixing depth thus should be determined by visual inspection with a sediment profile camera or core sample dating of ^{210}Pb. Lighter colored oxygenated zones maybe seen above darker anaerobic zones. Since the levels of ^{210}Pb decrease with time of sedimentation, it usually decreases with depth. The mixing of sediments, therefore, becomes homogenized throughout the various levels. The levels where ^{210}Pb are constant indicate the mixing level (Christensen, 1982). Mixing levels are typically up to 20 cm in depth (WDOE, 1995), but are typically from 5 to 15 cm (Figure 5.4). Highly contaminated aerobic sediments may have no biological activity and thus a 0-cm mixing depth.

Contaminated sediments can be carried downstream, thus causing a wider problem. Floods in rivers can be particularly problematic (Cardenas et al., 1995). Large storms in lakes and coastal regions can lead to significant sediment transport (Lick et al., 1994). Subsequently increased bioavailability, intake by organisms, and solubilization may occur.

5.3 EVALUATION OF THE NATURAL RECOVERY OF SEDIMENTS

As on land, the application of natural attenuation requires the understanding of the sediment–contaminant interactions, in addition to the environmental conditions. Much less information exists regarding the natural attenuation of heavy metals than for organic chemicals, although there are numerous partitioning mechanisms that can play a role. Dredging, however, can disrupt these conditions by increasing the oxidation conditions, which can lead to increased mobility and bioavailability of heavy metals. Zinc, copper, lead, cadmium, nickel, and mercury have all increased in mobility during dredging (Darby et al., 1996). In areas that are particularly sensitive to ecological damage when the sediments are disturbed due to an engineered remediation, MNR may be appropriate because a prime objective of MNR is to restore the habitat (Figure 5.5).

Multiple lines of evidence are needed for the evaluation of the potential for natural recovery of sediments (Figure 5.6).

- Documentation of the reduction in exposure and toxicity
- Identification of the primary reduction mechanisms and the rates of reduction
- Data for projection of the processes into the future

Methods for evaluating the rates of natural recovery include exponential time decay function data and dated sediment cores for determining contaminant transport. Some rates are shown in Table 5.1.

To evaluate the effectiveness of the natural recovery, the bioavailability, toxicity, and transport of the contaminants must be reduced (Figure 5.7). This in turn reduces risk. Source control must be completed before or with the recovery as part of a sustainable system. Both source loading and recovery processes in a complex system will require modeling for prediction of the future.

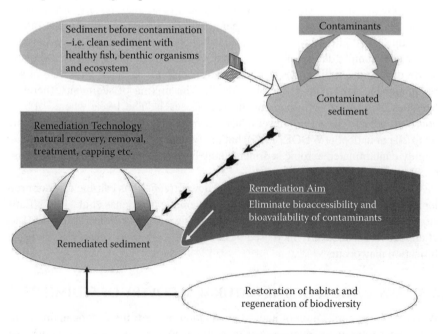

FIGURE 5.5   Objectives of natural recovery as a remediation process.

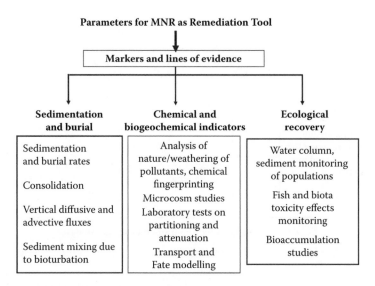

FIGURE 5.6   Lines of evidence indicators.

TABLE 5.1
Natural Recovery Contaminant Half-Lives in Sediments

Source	Contaminant	Half-Life (years)	Reference
Fox River, WI (DePere to Green Bay)	PCB	5.4	BBL, 1999
Hudson River, NY	PCB	10.4	QEA, 1999
Kalamazoo River, MI	PCB	6.7	BBL, 1999
Lake Anne (VA), Lake Blackshear (GA),	Chlordane	9.4	Van Metre et al., 1998
	DDT	12	
Lake Harding (GA),	Lead	9.8	
Lake Walter F. George (GA), Lake Seminole (FL/GA), White Rock Lake (TX), Coralville Reservoir (Iowa), Elephant Butte Reservoir (NM), Lowrence Creek (TX)	PCB	9.4	
Nassau Lake, NY	PCB	12	BBL, 1999
Lavaca Bay, TX	Mercury	3.2	Santschi et al., 1999
Netherlands	HCB[a]	7	Beurskens et al., 1993

[a] HCB—hexachlorobenzene.

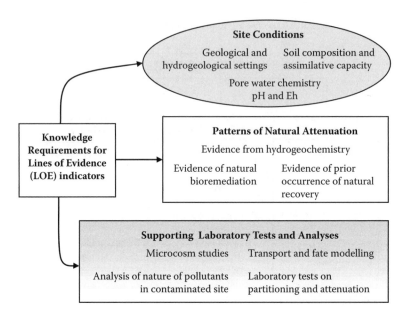

FIGURE 5.7 Required information and analyses for *lines of evidence* (LOE) indicators.

Short-term risk analyses need to be performed also. This includes risk to the public, workers, and the environment. Since there is no sediment handling in comparison to other management techniques such as excavation, dredging, sediment processing and transportation, risks are decreased.

Because there is no active remediation, costs are related to monitoring and potential institutional controls such as fish advisories. These are easy to estimate. It is more difficult, however, to estimate an acceptable time frame. The USEPA (1999) indicated for natural attenuation that a reasonable time frame estimation requires comparison of the alternatives for meeting the remediation objectives.

However, unlike on land, special environmental conditions in lakes, streams, rivers, estuaries, seas, and oceans can potentially lead to the mobilization of contaminants. Some of these include flowing water and currents which enhance mixing, dilution, and diffusion of contaminants. Storms and other high wave events may resuspend sediments. Bioturbation can cause particle mixing and solute transport, which can also influence pollutant movement. Knowledge on the presence and distribution of the infauna and their mechanisms of bioturbation will be required to predict the behavior of the contaminants in the sediments (Banta and Andersen, 2003). Natural capping through sedimentation of clean sediments can maintain the reducing conditions. Prediction of the sedimentation rates and contaminant fluxes can be highly uncertain.

Changes in pH, oxidative/reduction conditions, inorganic and organic complexation, and microbial populations can influence adsorption, absorption, sedimentation, and precipitation (Figure 5.8). The factors must be understood to determine the potential of natural attenuation for remediation of the contaminated sediments.

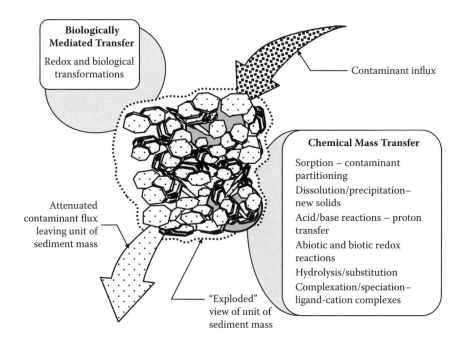

FIGURE 5.8 Processes involved in attenuation of pollutants in pore water through sediment (Adapted from Yong and Mulligan, 2004).

Techniques such as selective sequential extraction can assist in evaluating the potential for heavy metal mobilization.

For organic contaminants, knowledge of the k_{oc} and k_{ow} partition coefficients provides information on the contaminant characteristics. The strength of the bonding mechanisms and the age of the contamination in the sediments must be known.

Figure 5.9 shows the general protocol for considering monitored natural recovery (MNR) as a remediation of contaminated sediments. Site-specific data must be evaluated. Laboratory tests and predictive models are also necessary to provide information on the ability of the site materials and conditions to attenuate the contaminants. The Remediation Technologies Development Forum (RTDF) (Magar et al., 2004) has proposed that empirical methods for evaluating sediment stability include:

- Review of historical data
- Assessment of the geomorphology that includes landforms and their changes
- Measurement and modeling of hydrodynamic characteristics
- Evaluation of sediment erosion and transport
- Profiling the sediment core chemistry
- Survey of the hydrographic characteristics

Modeling would be necessary to evaluate the erosion potential at specific sites and transport and deposition of the sediment. Historical information must be site specific to validate the models. Flood, high waves, or tidal influences must be

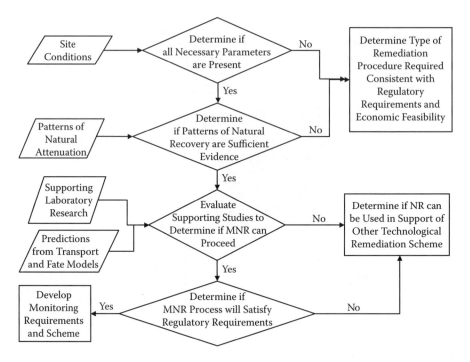

FIGURE 5.9 General protocol for considering MNR as a remediation tool (from Yong and Mulligan, 2004).

incorporated into the modeling evaluation. For geomorphological studies, sediment texture and thickness, hydrology/hydrodynamic data, and sediment transport data are used. Side-scan sonar, conventional probing, and sampling are useful in obtaining the data. Insufficient conditions for natural recovery would require technological remediation such as capping with clean sediments, sand, or other materials or other methods including dredging.

Monitoring is essential in determining the success of natural recovery in reaching the remediation goals and, thus, the use of the terminology, MNR. The pollutants in both the water and sediments need to be tested to ensure that the environmental conditions (pH, redox changes) have not changed and that the pollutants are not released into the environment. In addition, the laboratory tests and models used for prediction can be confirmed by the monitoring. Human and other receptors must be protected from exposure to the contaminants to reduce the potential for bioaccumulation and other damage. The costs can be significant but are the main aspect of the MNR. Monitoring has been promising at many sites (USEPA, 2005). There is a lack of information on fish tissue results in the long term.

Comparisons must be made to active remediation in regards to risk and impacts. Natural recovery may be possible in time periods that are not substantially longer and may have significant social and environmental benefits. Technologies such as dredging are not noninstrusive. There is also habitat destruction and the limitations of landfilling or other means of sediment disposal.

Once all the data is obtained, a weight-of-evidence was suggested by the RTDF (Magar et al., 2004) to balance the findings. The approach included:

- Identify the geomorphic areas with similar sediment and contaminant characteristics.
- Organize into a matrix or graphical format to present the data.
- Provide conclusions on long-term stability.
- Identify the implications of the predictions.
- Discuss uncertainties.
- State conclusions.

Performance-based natural recovery should be used along with contingency plans. If the natural attenuation is not performing according to the standards required, other remediation options then could be adopted. The standards should be based on effectiveness, implementability, and costs of the remediation technologies. NRC (1999) showed an overall assessment of the feasibility, effectiveness, practicality, and costs of various approaches. They ranked all aspects on a scale of 0 to 4. Natural recovery ranked as a concept that had no verification experimentally, that was commercially available, that was not very certain in terms of public acceptance, but was of very low cost (<$1/m^3). This comparison was very generalized, and decisions should be made on a site-specific basis. Additional considerations can include short-term releases, degree of control, risk reduction rate, and magnitude of the treatment long term risk. MNR generally is very low in risks on the short-term releases and treatment magnitude. However, the risk reduction rate and degree of control may be very unfavorable. Long-term risk can vary between low and high. Comparison to other remedies should include removal, enhanced NR, hydraulic modification, capping, and various combinations of technologies. Overall, MNR must achieve the remediation objectives in an adequate time with minimal risk to humans and the environment.

If the rate of MNR is not acceptable, then enhancement can be considered. A thin layer of sediment or sand can be employed to accelerate concentration reduction and achieve cleanup goals. Other materials such as carbon may also be added to reduce bioavailability.

5.4   MODELS FOR NATURAL REMEDIATION

Various models are used to incorporate the effect of the various mechanisms and to predict natural remediation (Figure 5.10). The types of models vary in complexity and are described by Dekker et al. (2007). The first level can include simple correlations and statistical models. The second level would be a conceptual model with the incorporation of various trends and observations. The next two levels (third and fourth) consist of numerical models. The third level is based on mass balances such as fluxes in and out of an area, rate of sediment accumulation, and so on. The fourth level is much more detailed and comprehensive. Selection of the appropriate level is usually based on the available data and resources and the type of information required. In general, the more complex the site, the more complex the model that is required. Model development for sediments usually will follow a sequence of simple then

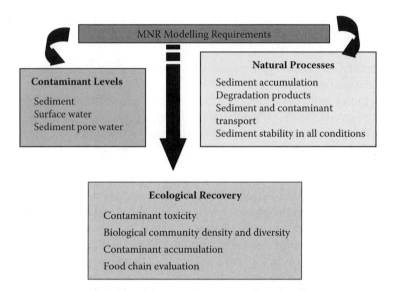

FIGURE 5.10   MNR modeling requirements.

detailed hydraulic models, then sediment transport models. Models for physical and chemical fate and transport are linked to the hydraulic and sediment transport models. Biological models including food chains and toxicity can also be incorporated.

A balance is needed between the modeling efforts and the information for decision making. The model should assist in identifying critical parameters in the system. Data used in the models should be based over extensive time periods, especially if the model is to be used for prediction of contaminant fate and transport. Calibration should be followed by verification. Long-term stability is required to ensure that recovery will be predicted accurately. If disturbances occur, they need to be incorporated into the model. Uncertainty will increase as the prediction period increases. Sensitivity analyses or probabilistic modeling can assist in identifying where the uncertainties are located or how far into the future the model can be realistically used.

Some available models include SEDCAM (WDOE, 1991), CORMIX, and WASP4 (Water Quality Analysis Simulation Program, developed by the USEPA). A newer one for evaluating the fate of sediments and their contaminants is EFDC (Environmental Fluid Dynamics Code). It has been used at Superfund sites such as Housatonic River, MA. TSS and PCB concentrations were simulated downstream at this site and were quite successful (Hayter, 2006).

Site-specific data is required to prove that the attenuation is occurring, in addition to calibrating and verifying fate and transport models for prediction of the future for the evaluation of sediment stability and subsequent risk to human and ecological receptors. Data requirements include deposition rates, sediment mixing layer thickness, and change in contaminant bioavailability and toxicity over time. Information on sediment contaminant distribution, hydraulics, hydrology, and sediment erosion, deposition, and transport is required for more complex models. Issues such as long-term evaluation of natural recovery for risk reduction and potential effects of extreme events such as floods are addressed by deterministic models.

Model predictions can be uncertain, are difficult to verify and calibrate, and the regulators may be skeptical, particularly due to the expertise and complexity involved. Therefore, selection criteria can include model openness and good documentation, perception by users (track record) and regulators, ability to test hypotheses, and ability to compare alternatives. Models should not be used as a replacement for data collection. Some of the data needs will be discussed in the following sections.

5.4.1 DEPOSITION RATE

Sediment deposition rate data is required to evaluate the potential for NR. Variability from site to site can be significant. Radioisotope dating measures ^{210}Pb and/or ^{137}Cs from sediment core samples. As the sediments are buried with new sediments, ^{210}Pb can be determined because it decays at a known rate. Its half-life is 22.3 years. Each section in the core sample can give the time scale. ^{137}Cs data can be also used with ^{210}Pb; unlike lead, ^{137}Cs is not a natural element. It is usually the result of nuclear testing. The maximum level was found in 1963 to 1965. Activities that disrupt the natural sedimentation rates can cause inaccuracies in the analysis. Some examples include dredging, dumping, natural settlement, and boat propellers. Dredging records can be consulted but may not be accurate. Sediment traps (glass cylinders of about 0.5 m in depth) can be used to collect settling particles (Norton and Michelsen, 1995). These particles may include new and resuspended materials. A combination of isotope dating and the use of sediment traps is often used to check the data as discussed in Chapter 4.

5.4.2 SOURCE LOADING

To determine the input of contaminants, the deposited sediments need to be analyzed for contaminants. This information with the deposition rate data will provide the influx of contaminants. The influence of resuspended sediments must be characterized to provide an accurate determination of the loading in the area. Initial concentrations of the contaminants at the surface and the variability over the area must also be determined.

5.4.3 HYDRODYNAMIC PARAMETERS

Models such as WASP4 require various hydrodynamic parameters (PTI Environmental Services, 1993). Some include bathymetry and shoreline orientation, location of the sources, stratification in the water column, currents (direction and magnitude), advection and dispersion rates, and particle size distribution and densities. This type of data usually would be needed for multiple sources and for complex sites such as rivers but may not be necessary at the screening stage.

Adsorption reactions or processes involving organic chemicals and soil fractions are governed by (a) the surface properties of the soil fractions, (b) the chemistry of the pore water, and (c) the chemical and physical–chemical properties of the pollutants. In general, organic chemical compounds develop mechanisms of interactions which are somewhat different from those given previously for inorganic contaminants. If the transport of organic chemicals in soils is considered, interactions between the contaminant and soil surfaces are important in predicting the retention capacity of the soil and the bioavailability of the contaminant. The interaction mechanisms are influenced by soil fractions, the type of and size of the organic molecule, and the presence of water. As in the case of inorganic contaminant–soil interaction, the existence of surface active fractions in the soil such as soil organic matter (SOM), amorphous noncrystalline materials, and clays can enhance contaminant retention in soils significantly because of large surface areas, high surface charges, and surface characteristics.

5.5   REGULATORY FRAMEWORK

In the decision-making process, natural recovery is usually considered as a cleanup alternative. It is not a "no action" alternative for several reasons. The ecological and human risks must be minimal and thus must be protected throughout the process. There can also be substantial costs for field studies, modeling, and monitoring to ensure that the risks are minimal and that objectives are being met. In addition a remedial contingency place must be conceived in the event that the objectives will not be met. Sediment quality guidelines can be used with modeling to determine the time frame and feasibility of MNR. As previously mentioned in Chapter 1, the State of Washington, USEPA, and Oregon Department of Environmental Quality have specifically mentioned the utilization of natural recovery in their strategies.

5.6   PROTOCOLS DEVELOPED FOR MONITORED NATURAL RECOVERY

Although previously guidance information was limited for sediments and concentrated mainly on soil/groundwater systems, the USEPA moved to clean up approximately 140 contaminated sediment sites. Thus guidance was established for MNR (USEPA, 2005). Site characteristics and conditions that would be amenable to MNR were listed as shown in Figure 5.11.

To evaluate MNR as a remediation option, ecological risk assessments are now required in North America and increasingly in Europe. Sediment Quality Guidelines (SQG) are used as part of the evaluation. Some of the aspects to complete the ecological risk assessments include (Apitz et al., 2005):

- Evaluation of the nature and extent of the contamination
- Determination of the indices of benthic diversity and numbers without contamination
- Evaluation of the bioavailability, bioaccumulation, and other effects on the organisms in the presence of the contaminants
- Evaluation of the fate and transport of the sediments and associated contaminants
- Evaluation of the potential risks of the contaminants toward the aquatic biota and surroundings

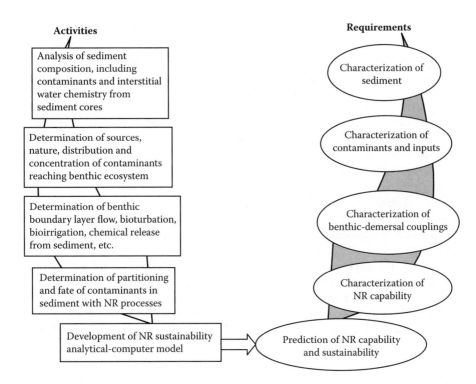

FIGURE 5.11 Activities and requirements for evaluating the capability and sustainability of NR.

Risks (short- and long-term) include habitat loss, impact on water quality, recreational activities, and aquatic life. On the long-term, the potential for recontamination needs to be evaluated, in addition to the recovery of the ecosystem. Meaningful goals must be established because ecological recovery will be slower than sediment recovery. Statistical variability will need to be monitored to determine if MNR is significant.

To evaluate natural recovery, various forms of evidence are needed including:

1. Natural contaminant burial and deposition of clean sediments
2. Evidence of sorption, precipitation, and other processes for reduction of contaminant mobility
3. Indication of biological or chemical transformation processes
4. Dispersion of particle-bound contaminants to reduce contaminant concentrations

Monitoring programs must include multiple lines of evidence (chemical, physical, biological, and geotechnical) in addition to modeling to evaluate the effectiveness of MNR. Risks of contaminant release must be minimized to achieve acceptable risk levels. Public perceptions can be enhanced by habitat restoration.

The majority of the available protocols address only fuel hydrocarbons or chlorinated solvents. Other organic contaminants such as PAHs, PCBs, explosives, and pesticides are not addressed, while metals, inorganics, and radionuclides are infrequently discussed. Therefore, there are major shortcomings in these protocols. One aspect not considered by the NRC, since their focus was groundwater, was that most protocols are designed for groundwater natural attenuation and not for soil natural attenuation or sediments. Few protocols exist for soil with the exception of the USEPA (1998) and those by the Department of Energy (DOE). Sediments have not received much attention at all. Sediments differ from land-based soils, since they can be transported and have higher organic contents. Oxidizing conditions only exist near the sediment surface. Sulfate reduction in marine sediments and methanogenesis in freshwater sediments tend to dominate.

Organisms can transport contaminants, and there is considerable variability at sites. Technical protocols have not been developed for sediments. The USEPA (2001) has recommended that research be expanded to determine natural attenuation mechanisms in sediments, monitoring methods need to be developed for quantifying natural attenuation, the contaminant transport, and bioaccumulation for analysis and assessment. Research specific for fresh water, coastal, and marine aquatic environments is also required. A protocol adapted from Yong and Mulligan (2004) for sediments can be seen in Figure 5.9.

To demonstrate that natural attenuation is taking place, lines (and possibly multiple lines) of evidence are established to indicate decreases in contaminant concentrations (USEPA, 2005). They include:

- Decreases in contaminant concentration in the higher-trophic-level biota (fish) over time
- Decreasing concentrations in the water column under low-flow (summer) conditions due to high biological activity
- Sediment core data showing decreasing trends in contaminant concentrations over time
- Decreasing trends in surface sediment contaminants and sediment toxicity of contaminant mass in the sediment

Spatial and temporal sediment characteristics may vary substantially and thus must be taken into account.

Numerous tools are available for field measurement to evaluate MNR. They include tools for physical transport measurement such as sedimentation, evaluating contaminant weathering, and evaluation of ecosystem impacts and recovery. Weathering processes include the physical, chemical, and biological processes that have already been discussed.

5.7 CASE STUDIES OF NATURAL RECOVERY

In this section, various case studies of natural recovery will be examined. Some examples are indicated in Table 5.2. The ability of sediments to retain organic contaminants is one of the keys to natural attenuation. At another site in Germany, a lignite seam accumulated aliphatic and aromatic chlorinated hydrocarbons downstream from a chemical plant (Dermietzel and Christoph, 2002). A two-compartment model

TABLE 5.2
Natural Recovery Sites

Location	Reason for MNR Selection and Results (if available)
Kapone, James River	Active remediation estimated between $3 to 10 billion and would disturb habitat
	Burial of sediments likely or dilution and mixing
Interstate Lead Company Superfund Site (AL), 1995 ROD	Historical declining trend in sediment lead concentration
	No evidence of ecosystem damage
	Active remediation would affect ecosystem
	Natural recovery would cause minimal disturbance
Lake Harwell Superfund Site (SC), 1994 ROD	Active remediation not practical or too costly
	Fishing advisories to reduce risk
	Source control at former Sangamo-Weston plant put in place
	1-D (HEC-6) model predicted a recovery rate of 1 mg/kg of PCBs which was reasonable
	1 mg/kg ROD surface cleanup goal achieved in 1999 after 20 years
Eagle Harbor (WA)	Source control and capping utilized
	Monitoring initiated
	Bathymetry
	Sediment grabs
	Cap coring
	Evidence of anaerobic oxidation
	Degradation rate of low MW PAHs > high MW PAHs
	Mobility of low MW PAHs reduced
	Acute toxicity of 2 and 3-ring PAH reduced
Bellingham Bay, Washington, USA	Discharges of Hg into bay (10-15 m in depth, >100 km2), monitoring and modeling assessment indicating MNR potential
	Sedimentation, erosion/resuspension, biological and physical mixing in the bed
	By 2002, surface sediments recovered to below Washington State standards

Source: Adapted from Magar (2004).

was shown to approximate the experimental results. An initial fast desorption based on transfer from the outer surface of the sediment was followed by a slower diffusion-controlled release from the interior of the sediment.

In 1998, sediment samples at Lake Harwell, SC, were taken at five places to determine the occurrence of natural attenuation of polychlorinated biphenyls (PCB) (Pakdeesusuk et al., 2005). The mole percentage of each congener of PCB and/or changes in the total of meta, para, and ortho chlorines and total chlorines per biphenyl were determined and compared to 1987 sediment samples. Solubilization and desorption were negligible according to mass balances. It was concluded that in situ dechlorination was occurring at a slow rate since 1987, after an initial rapid rate. Microcosm studies supported the findings. Lack of information on organic matter and electron acceptors such as nitrate, sulfate, iron, and manganese make it difficult to predict optimal dechlorination conditions. Capping with fresh sediment may need to be increased to decrease the risk of bioaccumulation in fish.

Trichloroethene contamination in the groundwater was first detected in 1982 at a Michigan National Priorities List site (An et al., 2004). Samples were monitored in 1991, 1992, 1994, 1995, and 1998, 100 m from the shore and later, 3 m from the shore. Products of dichloroethene (DCE), vinyl chloride (VC), ethene, and methane were found, indicating anaerobic degradation. Degradation rates were estimated using a two-dimensional (2D) model. Analysis of the water in the lake sediments indicated natural attenuation.

Although most protocols indicate the dominance of the biological degradation, other mechanisms may also be significant. Ferrey et al. (2004) indicated that, although there was no evidence of biodegradation, *cis*-dichloroethylene (*cis*-DCE) and 1,1-DCE, iron minerals such as magnetite removed these compounds from sterilized sediments by reductive dechlorination. Sorption, particularly to organic matter, did not appear to be responsible for the loss.

At the Columbus Air Force Base, Columbus, MS, 60 sediment samples were taken to evaluate the fate and transport of jet fuel contaminants (Stapleton and Sayler, 1998). DNA probes were used to determine the genes for the following degradative enzymes: alkane dioxygenase, toluene monooxygenase, naphthalene dioxygenase, toluene dioxygenase, toluene monooxygenase, xylene monooxygenase, carbon monoxide dehydrogenase, and methyl coenzyme reductase. 10^7 to 10^8 organisms per gram of sediment were found, compared to 10^4 to 10^6 organisms per gram by traditional methods. Degradation of BTEX and naphthalene were also indicated, particularly after five to seven days. More than 40% of these ^{14}C-labeled compounds were mineralized in the sediments, without nutrient addition. Correlations of laboratory assay and field analyses are required, and thus further field tests will be performed.

At the Dover Air Force Base, Dover, DE, contaminated with chlorinated ethenes, a characterization of the microbial community was performed (Davis et al., 2002). Low biomass levels (10^7 bacteria per gram sediment) were found. Mineralization of vinyl chloride and *cis*-DCE was occurring, and 16S rRNA gene sequence indicated the presence of anaerobic microorganisms that were capable of anaerobic halorespiration and iron reduction. The data showed that the microorganisms were the major mechanism for reductive and oxidative attenuation of the chlorinated ethenes.

The weathering of PAH-contaminated sediments was monitored by Brenner et al. (2002) at the Wyckoff/Eagle Harbor Superfund Site near Seattle, WA. Three PAH sources were determined: creosote, urban runoff, and natural background. Urban runoff was found to contribute to the contamination over the past 50 to 70 years. Unweathered and pure-phase creosote deposits were found below 30 cm in depth. However, surface sediments (upper 20 to 30 cm) were a mixture of weathered creosote and urban runoff. Lower-molecular-weight PAHs in particular were lost in creosote-contaminated weathered sediments. Capping of 1 to 3 m of clean sand was performed because the deposit of clean sediments was not extensive due to continuous contamination from urban runoff.

Moser et al. (2003) evaluated a "freeze core" sampling method for determining the geochemistry and microbiology of sediments contaminated with chromium (VI).

Liquid nitrogen was used to freeze the cores. Significant numbers of sulfate, nitrate, and iron-reducing bacteria in addition to amounts of acid-volatile sulfide were found, but the freezing decreased the numbers of viable bacteria. This indicated the potential for a combination of anaerobic microbial and chemical processes to contribute to the natural attenuation of chromium at the Hanford site. Reduction of chromium (VI) to chromium (III) decreases its solubility and toxicity.

Arias et al. (2003) also studied Cr(VI) natural attenuation in sediments by laboratory mesocosms to mimic environmental conditions. Cr accumulated in the upper 5 mm of the sandy sediments. However, Fe, Mn, and total organic contents did not correlate with total Cr levels. PCR of 16S rRNA genes were used to analyze microbial populations and indicated that the microbial population were inhibited and therefore provided little information regarding the bacteria present if natural attenuation is to be employed.

In the Hackensack River, sediment was contaminated by chromium due to deposal of 800,000 m^3 of chromite ore processing residue during the period of 1905 to 1954. Because chromium (III) is stable, monitored natural recovery (Evison et al., 2007) was implemented because the form is not likely to oxidize to Cr(VI) even with dissolved oxygen.

A study of heavy metals was conducted in Port Philip Bay, Australia, because there were indications of toxic metals in fish and shellfish (Fabris et al. 1999). The objective was to determine the partitioning of heavy metals in dissolved and particulate species in the bay waters. Despite a flushing time of 10 to 16 months in the bay, concentrations in the near shore and estuarine areas were not higher than those in the coastal marine waters. Most of the mechanisms for partitioning were related to precipitation of iron and manganese oxyhydroxides that coprecipitate with dissolved heavy metals. There was a strong correlation of iron with chromium, nickel, and zinc in the particulates. Contrary to the metals, arsenic concentrations (As(III)) increased in depth in the sediments and thus did not seem to be the result of anthropogenic activity. Near the surface layer of sediments, arsenic is oxidized to As(V) and leaves the sediments. Fe(III) can coprecipitate some of the arsenic and become trapped in the sediments.

Sampling over time often indicated the occurrence of NR. In the Bellingham Bay, WA, area, Hg was followed over time. In the area of Nooksack River, there was a substantial source of clean sediments (Hart Crowser, 1997). The trends thus showed that the remediation was probably occurring in this area at a sufficient rate.

The recovery of aquatic ecosystems can be remarkable over time. In the Great Lakes Basin, human and ecological targets have been impacted over time. There have been many commercial and sports fishing advisories, and dredging has been limited due to disposal issues. Wildlife has been impacted, eutrophication has increased, and populations of fish have various deformities. MNR is attractive in this area because it is potentially less costly and disruptive. Monitoring has been very difficult to quantify due to the lack of planning of a monitoring program. The by-products have not been easily detected and monitored. This is needed to distinguish natural bioremediation and other physical processes such as volatilization, migration and sorption. Pesticides have attenuated over time (Brown, 1999). Attenuation of other contaminants such as DDT and PCBs has not been as extensive. However,

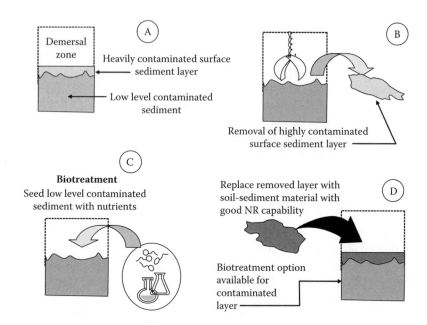

FIGURE 5.12 Enhancement of bioremediation for NR.

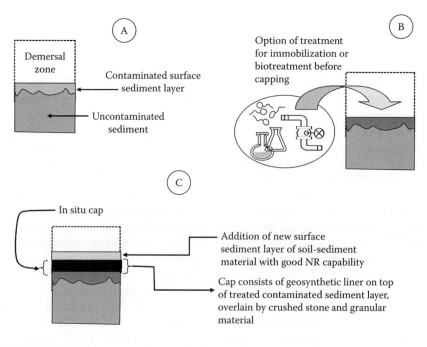

FIGURE 5.13 Enhancement of contaminant retention processes of NR.

the cormorant, peregrine falcon, and bald eagle populations have recovered notably. PCBs and DDT levels in lake trout have reduced by half every four to eight years (De Vault et al., 1996). More than 150 case studies were documented by Niemi et al. (1990) of processes for restoring freshwater ecosystems.

A particular case involves the St. Clair River Area of Concern between Lake Michigan and Lake Huron (Passino-Reader et al., 2000). It has received pollution from municipal sewage treatment plant effluents, industrial wastes, and other discharges. The contaminants included heavy metals, petroleum compounds, phenols, chlorinated organics, and suspended solids. Highly contaminated sediment hot spots have been found. Various impairments resulted from the contamination such as fish and wildlife consumption, beach access, drinking water restriction, and aesthetics. Due to the complexity of the site, many modifications were needed to control the contaminant sources, and many parties were involved. Although NR was considered at the site, it was not adopted immediately and would be further discussed later (USEPA, 1998). Discussion among all parties was necessary. Some of the recommended changes involved the upgrading of the wastewater treatment and waste management facilities, construction of a wetland modifying industrial and agricultural practices, and dredging in selected areas.

5.8 ENHANCED NATURAL RECOVERY

Capping of thin layers of sand (e.g., 10 to 15 cm) is known as enhanced natural recovery (ENR). The grain size and organic content of the capping material must be considered carefully, and water programs must be consulted before introducing new materials into a water body. Natural materials are thus preferable. Erosion of this material will need to be considered in sensitive areas such as shellfish beds. The thin cap helps to isolate the sediment on the short term from the benthic organisms to create a clean surface material. It also assists in the remediation by bioturbation through dilution and mixing of the contaminated sediments with clean sediment. Degradation processes are also enhanced. Other materials such as activated carbon could be added to enhance the bioremediation (Figure 5.12) or the sorption capability of contaminants (Figure 5.13).

Natural recovery may also complement other remediation techniques. At the Sitcum Waterway of the Commencement Bay estuary, dredging was performed in 1993–1994 to remove heavy metals, PAHs, and PCBs. Although the removal was successful overall, there was a portion of the side slope that still exceeded sediment quality objectives. MNR was evaluated at this location. Site investigation included the use of sediment traps, sediment core profiling and radioisotope analysis, and simple recovery models. It was determined that the NR would be accelerated due to the dredging. Sediment quality monitored was then implemented in 1998 and 2003. Lead and high-molecular-weight PAHs were reduced to below the required standards as predicted by the model. Acceleration of the NR was achieved by resuspension of clean sediments that redeposited onto the sediment, reducing the concentrations (Patmont et al., 2004). In addition, MNR can be considered for dredge residuals.

At a site in the Lower Duwamish Waterway, Seattle, WA, dredging was performed in 2004 (Stern and Colton, 2009). An adjacent area indicated higher levels

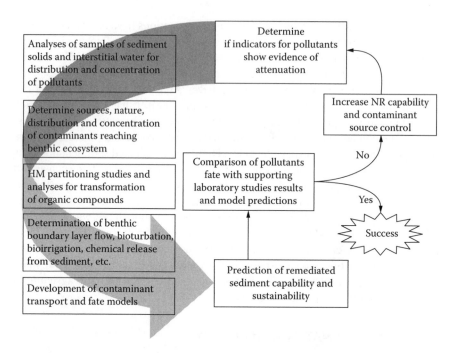

FIGURE 5.14 General protocol to determine feasibility and application of NR as a tool for impact mitigation and management.

of PCBs. It was predicted than MNR was 2 to 3 years without further remediation. A thin layer of clean sand (less than 0.3 m) was placed for ENR. The levels were immediately reduced to predredge levels. The choice between MNR and ENR will depend on the time frame to meet the cleanup objectives and risk considerations and sedimentation rates.

5.9 CONCLUDING REMARKS

There are various advantages regarding the natural attenuation of sediments including the possibility of destroying the contaminant completely, potential for reduction in remediation costs, and limited site disruption. Its main application is for areas of low-level contamination if used as the main approach. However, it may also be utilized after a remediation action has occurred. Potential disadvantages, however, include longer remediation times compared to other technologies, a lack of knowledge concerning mechanisms for remediation, particularly with regard to inorganic contaminants, substantial requirements for monitoring, and the potential for desorption or resolubilization of contaminants.

Although there are many similarities between existing protocols and guidelines for natural attenuation, particularly concerning lines of evidence and data requirements, very few consider the sediments in their protocols, and most have been adapted for hydrocarbon and chlorinated solvent contamination. The natural attenuation of many other contaminants has been limited for PAHs, PCBs, pesticides, and inorganic contaminants. The interaction of organic and inorganic contaminants with the soil components may also be an important factor in natural attenuation processes. Some of the natural attenuation processes that are specific for sediments are sediment deposition and resuspension, mixing due to wave action, and bioturbation and thus are not included in groundwater protocols. Monitoring is the key for documenting and evaluating the success of the sustainability of MNR (Figure 5.14).

In addition, there is little information available regarding the comparison of the natural recovery processes in the various sediment environments (rivers, estuaries, lakes, and coastal seas and oceans). The processes are site specific and therefore, the evaluation of these processes must also be.

Indicators are needed to evaluate the risk of contaminated sediments and to monitor and assess the progress of MNR. Biological indicators to evaluate the ecosystem status are particularly needed. PCB persistence in sediments is particularly problematic.

REFERENCES

An, Y.-J., Kampbell, D.H., Weaver, J.W., Wilson, J.T., and Jeong, S.-W. 2004. Natural attenuation of trichloroethene and its degradation products at a lake-shore site. *Environ. Pollut.* 130: 325–335.

Apitz, S.E., Davis, J.W., Finkelstein, K., Hohreiter, D.L., Hoke, R., Jensen, R.H., Kirtay, V.J., Jersac, J., Mack, E.E., Magar, V., Moore, D., Reible, D., and Stahl, R. 2005. Assessing and managing contaminated sediments: Part I: Developing an effective investigation and risk evaluation strategy. *Integr Environ. Assess Manag.* 1: 2–8.

Arias, Y. M., Obraztsova, A., and Green-Ruiz, C. 2003. Natural attenuation of Cr(VI) contamination in laboratory mesocosms. *Geomicrobiol. J.* 20: 389–401

Banta, G. and Andersen, O. 2003. Bioturbation and the fate of sediment pollutants—Experimental case studies of selected infauna species. *Vie et Mileu.* 53: 233–248.

BBL. 1999. Draft—Loeffel Site Environs Revised Study Report: Nassau Lake Drainage Basin. Syracuse, New York.

Beurskens, J.E.M. Dekker, C.G.C., Jonkhoff, J. and Pompstra, L. 1993. Microbial dechlorination of hexachlorobenzene in a sedimentation area of the Rhine River. *Biogeochemistry* 19: 61–81.

Brenner, R.C., Magar, V.S., Ickes, J.A., Abbott, J.E., Stout, S.A., Crecelius, E.A., and Bingler, L.S. 2002. Characterization and FATE of PAH-contaminated sediments at the Wyckoff/Eagle Harbor Superfund site. *Environ. Sci. Technol.* 36: 2605–2613.

Brown, M.P. 1999. The role of natural attenuation/recovery processes in managing contaminated sediments. BBL, Fall.

Callendar, E. 2005. Heavy metals in the environment—Historical trends. In B.S. Lollar (Ed.), *Environmental Geochemistry*, Elsevier, Amsterdam.

Cardenas, M., Gailani, J., Ziegler, C.K. and Lick, W. 1995. Sediment transport in the Lower Saginaw River. *Mar. Freshwater Res.* 46: 337–347.

Cardenas, M. and Lick, W. 1996. Modeling the transport of sediments and hydrophobic contaminants in the Lower Saginaw River. *J. Great Lakes Res.* 22: 669–682.

Christensen, E.R. 1982. A model for radionuclides in sediments influenced by mixing and compaction. *J. Geophys. Res.* 87: 566–572.

Darby, D.A., Adams, D.D., and Nivens, W.T. 1996. Early sediment changes and element mobilization in a man-made estuary marsh. In P.G. Sly (Ed.), *Sediment and Water Interactions*, Springer, Berlin, pp. 343–351.

Davis, J.W., Odom, J.M., DeWeerd, K.A., Stahl, D.A., Fishbain, S.S., West, R.J., Klecka, G.M., and DeCarolis, J.G. 2002. Natural attenuation of chlorinated solvents at Area 6, Dover Air Force Base: Characterization of microbial structure. *J. Contam. Hydrol.* 57: 41–59.

Dekker, T., Lautenbach, D., Magar, V., Pekala, J., and Wenning, R.J. 2007. A data-based and modeling assessment of hydrodynamics, sedimentation and sediment stability at the Hackensack/Passaic/upper Newark Bay confluence. Abstract B-004, Sediment Remediation, *Fourth International Battelle Conference on Remediation of Contaminated Sediments.* Jan. 22–25, Savannah, Georgia.

Dermietzel, J. and Christoph, G. 2002. The release of pollutants from aged field sediments. IAHS Publications, *IAHS Publications—Series of Publications and Reports, Internal. Assoc. Hydrol. Sci.* 275: 147–151.

DEQ Oregon Department of Environmental Quality. 1998. Contaminated sediment management strategy. 9/10/98 Revised Draft. Oregon Department of Environmental Quality, Waste Management and Cleanup Division, Portland, OR.

De Vault, D.S., Hesselberg, R., Rodgers, P.W., and Feist, T.J. 1996. Contaminant trends in Lake Trout and Walleye from the Laurentian Great Lakes. *J. Great Lake Res.* 22(4): 884–895.

Evison, L., Greenberg, M., Logan, M., Magar, V., Nadeau, S., and Reible, D. 2007. Definition and demonstration of remedy effectiveness. What worked and what didn't. (Panel Discussion) Panel 002, Sediment Remediation, *Fourth International Battelle Conference on Remediation of Contaminated Sediments.* Jan. 22–25, Savannah, Georgia.14 pp,

Fabris, G.J., Monahan, C., and Batley, G.E. 1999. Heavy metals in waters and sediments of Port Philip, Australia. *Mar. Freshwater Res.*, 50: 503–513.

Ferrey, M.L., Wilkin, R.T., Ford, R.C., and Wilson, J.T. 2004. Nonbiological removal of *cis*-dichloroethylene and 1,1-dichloroethylene in aquifer sediment containing magnetite. *Environ. Sci. Technol.*, 38: 1746–1752.

Hart Crowser. 1997. Remedial investigation report, Whatcom Waterway site, Bellingham, Washington. Draft report. Hart Crowser, Inc., prepared for Georgia Pacific West Inc. Bellingham, WA.

Hayter, E.J. 2006. Evaluation of the state-of-the-art contaminated sediment transport and fate modeling system. USEPA, Office of Research and Development, EPA/600/R-06/108, September, Washington, DC.

Lick, W., Lick, J., and Ziegler, C.K. 1994. The resuspension and transport of fine-grained sediments in Lake Erie. *J. Great Lakes Res* 20: 599–612.

Magar, V.S., Davis, J., Dekker, T., Erickson, M., Matey, D., Patmont, C., Swindoll, M., Brenner, R., and Zeller, C. 2004. Characterization of fate and transport processes: Comparing contaminant recovery with biological endpoint trends. Working draft. Remediation Technologies Development Forum (RTDF), Sediment Remediation Action Team.

Moser, D.P., Fredrickson, J.K., Geist, D.R., Arntzen, E.V., Peacock, A.D., Li, S.-M.W., Spadoni, T., and Mckinley, J.P. 2003. Biogeochemical processes and microbial characteristics across groundwater-surface water boundaries of the Hanford Reach of the Columbia River. *Environ. Sci. Technol.*, 37: 5127–5134.

Niemi, G.J., Devore, P., Detenbeck, N., Taylor, D., Lima, A. Pastor, J. Yount, D. J., and Naiman, R.J. 1990. Overview of case studies on recovery of aquatic systems from disturbance. *Environ. Manag.* 14: 547–587.

Norton, D. and Michelsen T.C. 1995. Elliott Bay waterfront recontamination study. Pub. No. 95-335. Washington Department of Ecology, Environmental Investigations and Laboratory Services, Olympia, WA.

NRC [National Research Council]. 1997. *Contaminated Sediments in Ports and Waterways. Cleanup Strategies and Technologies.* National Academic Press, Washington, DC.

NRC [National Research Council]. 1999. National Symposium on Contaminated Sediments, Coupling Risk Reduction and Sustainable Management and Reuse. Proceedings of a Conference May 27–29, 1998, Washington, DC. National Academy Press, Washington, DC.

NRC [National Research Council]. 2007. *Sediment Dredging at Superfund Megasites: Assessing the Effectiveness.* National Academies Press, Washington, DC, 236 pp.

PTI Environmental Services. 1993. WASP application guidance manual. Washington Department of Ecology, Olympia, WA.

Pakdeesusuk, E., Lee, C.M., Coates, J.T., and Freedman, D.L. 2005. Assessment of natural attenuation via in situ reductive dechlorination of polychlorinated biphenyls in sediments of the Twelve Mile Creek Arm of Lake Hartwell, SC. *Environ. Sci. Technol.* 39: 945–952.

Passino-Reader, D.R., Kamrin, M.A., and Hickey, J.P. 2000. Natural remediation in Great Lakes. In M. Swindoll, R.G. Stahl, Jr., and S.J. Ells (Eds.), *Natural Remediation of Environmental Contaminants: Its Role in Ecological Risk Assessment and Risk Management.* SETAC Press, Pensacola, FL, pp. 365–409.

Patmont, C., Davis, J., Dekker, T., Erickson, M., Magar, V., and Swindoll, M. 2004. Natural recovery: Monitoring declines in sediment chemical concentrations and biological endpoints. Working draft. Remediation Technologies Development Forum (RTDF), Sediment Remediation Action Team.

QEA. 1999. PCBs in the Upper Hudson River, Volume 2, A Model of PCB Fate, Transport and Bioaccumulation, Montvale, NJ.

Reible, D. 2006. Natural attenuation of contaminated sediments, Sediment Remedies: Monitored Natural Recovery—Technical Considerations for Evaluation and Implementation, U.S. EPA, Office of Superfund Remediation and Technology Innovation, Oct. October 30, 2006, Internet Seminar, http://www.clu-in.org/conf/tio/sedmnr_103006/.

Santschi, P.H., Allison, M.A., Asbill, S., Perlet, B., Cappellino, S., Dobbs, C., and McShea, L. 1999. Sediment transport and Hg recovery in Lavaca Bay, as evaluated from radionuclide and Hg distributions. *Environ. Sci. Technol.* 33: 378–391.

Stapleton, R.D. and Sayler, G.S. 1998. Assessment of the microbiological potential for the natural attenuation of petroleum hydrocarbons in a shallow aquifer system. *Microb. Ecol.* 36: 349–361.

Stern, J.H. and Colton, J.A. 2009. Comparison of Monitored Natural Recovery (MNR) and Enhanced Natural Recovery (ENR) effectiveness. Fifth International Conference on Remediation of Contaminated Sediments (Feb. 2–5, 2009), Jacksonsville, FL.

USEPA. 1994. EPA's contaminated sediment management strategy. U.S. Environmental Protection Agency, Office of Water, Washington, DC.

USEPA. 1995. Cleaning up contaminated sediments: A citizen's guide. Assessment and remediation of contaminated sediment (ARCS) program. U.S. Environmental Protection Agency, Great Lakes National Program Office, Chicago, IL. EPA 905/K-95/001. July.

USEPA. 1998. EPA's Contaminated Sediment Management Strategy, EPA823-R-98-001, U.S. Environmental Protection Agency, Office of Water, Washington, D.C.

USEPA. 1999. Use of monitored natural attenuation at Superfund, RCRA Corrective Action and Underground storage tank sites. U.S. Environmental Protection Agency, Office of Solid Waste and Emergency Response. Washington, DC. EPA 540/R-99/009, April.

USEPA. 2001. Science Advisory Board, Monitored Natural Attenuation; USEPA Research Program—An EPA Science Advisory Board Review. U.S. Environmental Protection Agency, Science Advisory Board (1400A) Washington, DC.

USEPA. 2005. Contaminated Sediment Remediation Guidance for Hazardous Waste Sites EPA540-R-05-012. U.S. Environmental Protection Agency, Office of Solid Waste and Emergency Response, December 2005.

Van Metre, P.C. 1998. Similar rates of decrease of persistent, hydrophobic, and particle reac-
tive contaminants in riverine systems. *Environ. Sci. Technol.* 32(21): 3312–3317.
WDOE. 1991. Sediment cleanup standards user's manual. Washington Department of Ecology,
Sediment Management Unit, Olympia, WA.
WDOE. 1995. Sediment sampling and analysis plan: Appendix: Guidance on the development
of sediment sampling and analysis plants meeting the requirements of the Sediment
Management Standards: Chapter 173–204 WAC. Washington Department of Ecology,
Sediment Management Unit, Olympia, WA.
Wenning, R.J., Sorensen, M., Leitman, P., and Magar, V. 2007. Risk of remedy consider-
ations in sediment remediation projects. Abstract A-034. Fourth International Battelle
Conference on Remediation of Contaminated Sediments, Savannah, Georgia, Jan.
22–25, 2007.
Yong, R.N. and Mulligan, C.N. 2004. *Natural Attenuation of Contaminants in Soils.* CRC
Press, Boca Raton, FL.

6 In Situ Remediation and Management of Contaminated Sediments

6.1 INTRODUCTION

The remediation of sediments may be required when the sediment leads to the accumulation of contaminants in aquatic life or when the release of hazardous materials from sediments becomes a serious problem. For example, the sediment quality criterion for this threshold point may be described as the frequent effective level (e.g., FEL, Environment Canada and Ministère du Développement Durable, de l'Environnement et des Parcs du Québec, 2007). The frequent effective level (FEL) can be defined as the concentration of the element or compound that frequently gives toxic effects to aquatic life. Beyond the FEL, aquatic life cannot appropriately be protected. Therefore, a remediation technique, such as capping, dredging, or physical, biological and/or chemical treatment, has to be considered. In this chapter, in situ remediation techniques and the management of contaminated sediments will be described. Frequently, only about 10% of the sediments in an area are contaminated. Thus in situ remediation could be beneficial over dredging due to a reduction in costs and the need for solid disposal. It could also potentially be a more permanent solution. Therefore, there is a growing acceptance that large volumes in particular can be managed through in situ techniques (Forstner and Apitz, 2007).

6.2 IN SITU CAPPING

Due to the submerged conditions, there are limitations to performing in situ remediation of sediments. The main purposes for the in situ remediation of sediments are:

1. To immobilize contaminants or contaminated sediments
2. To reduce or cease the release of contaminants from sediments
3. To extract contaminants from the sediments

For the first two objectives, in situ capping may be ideal and has been used for many years. A list of 109 capping projects has been published by the Hazardous Substance Research Centers (HSRC) (http://www.sediments.org/capsummary.pdf).

In situ capping (ISC) is defined as the placement of a subaqueous covering or cap of clean or suitable isolating material over an in situ deposit of contaminated sediment. This material can include sediment, soil, or sand. Geosynthetic materials are also utilized. Placement methods include using barges, hopper dredges, pipelines, or direct placement. The material must be placed in a uniform and accurate manner. ISC is a potentially economical and effective approach for remediation of contaminated sediment (Palermo, 1998). A number of sites have been remediated by in situ capping operations worldwide, because the technique is the least expensive remediation alternative available for marine sediments that are unsuitable for open water disposal. The U.S. Army Corps of Engineers (USACE) has developed detailed guidelines for planning, designing, constructing, and monitoring in situ capping projects for the United States Environmental Protection Agency (USEPA) (http://www.epa.gov/glnpo/sediment/iscmain/). In this report, case histories of in situ capping (six cases in Japan, one case in Norway, and one case in the United States) are included in the Appendix.

The design objectives of a cap are normally one or more of the following (Reible, 2005):

1. Physical containment of the underlying contaminated sediment
2. Separation of the contaminants from biota at the sediment–water interface
3. Isolation of the chemical contaminants from the overlying water
4. Encouragement of habitat values of the surficial sediments

Cap material may also be used to encourage adsorption, absorption, precipitation, and ion exchange mechanisms of the contaminants in the sediments.

Azcue et al. (1998) presented data from a demonstration in situ capping site (100 m × 100 m) in Hamilton Harbour, Lake Ontario, Canada. A layer of clean medium to coarse sand with an average thickness of 35 cm was placed at the site in the summer of 1995. Concentrations of Zn, Cr, and Cd in the original sediments reached values of over 6000, 300, and 15 mg/kg, respectively. In general, the concentrations of elements were greater in the pore water than in the overlying water (e.g., the concentrations of Fe and soluble reactive phosphorus were 1000 times, and those of Mn 100 times greater). There was a significant reduction in the vertical fluxes of all the trace elements after the capping of the contaminated sediments.

On the other hand, it is argued that the effects of a cap decrease with time and that, after a few years, the effects are not significant. It is likely that newly settled contaminated sediments will release contaminants, and that the cap can be gradually eroded by water currents and waves. Since the installed cap materials change the geographical aspects, restoration can occur as a result of drifting sand. The consolidation of sediments, induced by the weight of cap, will cause adverse effects (i.e., release of contaminated pore water from the sediments).

Mohan et al. (2000) presented the design principles and theoretical basis for underwater caps for isolating contaminated marine sediments. They described that the cap consists of a base stabilizing layer, an isolation layer, a filter layer, and an armor layer. In addition, they pointed out the following technical considerations for the required design including:

1. Consolidation analyses
2. Contaminant release mechanisms (diffusion, advection/dispersion, and pore water release)
3. Hydraulic analyses (maximum velocities from river flow, waves, propeller wash, and ice scour)

Other considerations that will need to be monitored include the ability of the cap to accommodate bioturbation and to accumulate contaminants. These principles will help the researcher/engineer to better understand the behavior of contaminants within a cap system, thereby enabling them to evaluate the effectiveness of underwater capping as a potential remedial measure for contaminated aquatic sites.

Moo-Young et al. (2001) also pointed out that advection–dispersion was the dominant transport process based on the centrifuge test. Berg et al. (2004) analyzed a calcite barrier capping to immobilize phosphorus in eutrophic lake sediments. The results showed that calcite barriers could be optimized in accordance with the hydrochemical conditions in lakes to increase the efficiency of P retention in sediments. Various models are available for the U.S. Army Corps of Engineers to model cap placement including STFATE, MDFATE, LTFATE, CDFATE, SSFATE, and DREDGE. Thus, there may be several advantages and disadvantages for utilizing capping techniques in lake and marine areas.

The main advantages are the possible isolation of contaminated sediments from living organisms, the low cost, and simplicity. On the other hand, the disadvantage is that the contaminants in the pore water can be discharged into the water column due to advection and diffusion. Advection will occur as a result of consolidation under a load of capping materials. The diffusion may be the subsequent stage when the consolidation is terminated. These processes will be promoted by the installation of a cap.

Some studies have been made to cover the disadvantages of a sand cap. They focused on adsorbents as capping materials (Jacobs and Förstner, 1999). Furthermore, reactive mats can be recommended as capping materials (Olsta et al., 2006).

6.2.1 Design Factors for Sand Capping

6.2.1.1 Consolidation

There may be two mechanisms in the consolidation of sediments. A change volume of the natural sediments may occur by an increase in the self-weight due to newly fallen suspended solids (SS). This mechanism can be called self-weight consolidation. Another consolidation mechanism can be explained by Darcy's law, where water drainage occurs due to the excess pore water pressure induced by the quick loading. The latter is well understood as primary compression and explained by the Terzaghi's consolidation theory (Terzaghi and Peck, 1967).

The sediments are formed by natural consolidation, which can be explained as the settling of particles with interactions. Herein, no excess pore water pressure is required for the natural consolidation process. This can be identified as secondary compression followed by primary compression under loading.

Cap materials act as a load on the contaminated sediments. However, the top layer of fine sediments are usually so soft that any load cannot be supported by the skeleton of particles, as indicated in Chapter 2. When the capping materials are applied to the surface of the sediments, the top layer will escape or mix with the cap materials. This very soft layer is often called the drifting layer and is about 5 cm in thickness (Fukue et al., 1987).

Near the surface, the water content of sediments will decrease rapidly with depth, because of the weak structure under the self-weight. If the load is applied on the sediments, excess pore water pressure will occur in the sediments. The pore water will move from the sediment to the cap materials due to the difference in water pressure between the sediments and cap materials. Darcy's law can be applied for this water flow. As a result, the settlement of the sediment occurs. This phenomenon is called consolidation and is used in geotechnical engineering to predict the settlement of soils.

Models have been developed to predict the advection and diffusion of contaminants, such as phosphorus, tertbutyl tin (TBT), or heavy metals from sediments to the cap materials and water column (Arega and Hayter, 2008; Hamer and Karius, 2005; Schauser et al., 2004). Their predictions using the models showed good results.

When capping materials are installed successfully, the consolidation load σ_c' is given by

$$\sigma_c' = \gamma_c' H_c \tag{6.1}$$

where γ_c', is the submerged unit weight of the capping materials, and Hc is the thickness of the cap. With this load, the consolidation occurs. The γ_c' of sand is around 10 kN/m^3. When H_c is 0.5 m, σ_c' becomes about 5 kN/m^2. The γ_c' is determined using the following relationship:

$$\gamma_c' = \frac{G_s - 1}{1 + e}\gamma_w = \gamma_{sat} - \gamma_w \tag{6.2}$$

$$e = \frac{G_s w}{100} \tag{6.3}$$

where G_s is the specific gravity of cap solids, e is the void ratio, and γ_{sat} is the submerged unit weight, w is the water content of cap materials, and γ_w is the unit weight of the pore water.

The properties of the sediment change with depth (Fukue et al., 1986). If the sediment is homogeneous with depth, the void ratio of sediments may decrease by 50% at a depth of about 5 cm from the top surface and continue to decrease with depth due to consolidation.

From consolidation tests, the parameters, the coefficient of consolidation C_v and compression index C_c, can be determined. The C_v is used to predict consolidation speed, and C_c is used to predict the settlement. The C_c is defined as the slope of the

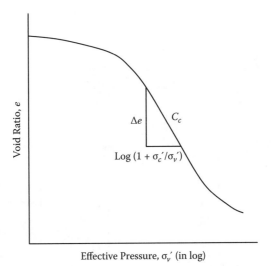

FIGURE 6.1 Void ratio–effective pressure relationship (*e*-log*p* curve) and the definition of compressive index *Cc*.

straight part of the void ratio–effective pressure relationship, as shown in Figure 6.1. Then,

$$Cc = \frac{\Delta e}{\log_{10}(1 + \sigma_c{'}/\sigma_v{'})} \tag{6.4}$$

where $\sigma_v{'}$ is the overburdened effective pressure of the sediments.

Theoretically, the settlement S with a thickness H is given by

$$S = \int_0^H \varepsilon_v(z)dz \tag{6.5}$$

On the other hand, strain ε_v can be expressed by

$$\varepsilon_v = \frac{\Delta e}{1 + e_0} \tag{6.6}$$

where Δe is the decrease in void ratio, and e_0 is the initial void ratio.

From Equations (6.4) and (6.6),

$$S = \int_0^H \varepsilon_v(z)dz = \int_0^H \frac{\Delta e}{1 + e_0}dz \tag{6.7}$$

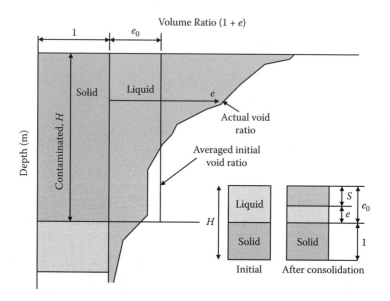

FIGURE 6.2   Illustration of void ratio for shallow sediment.

Since the void ratio profile changes with depth, Equation (6.7) (Figure 6.2) can be applied to small slices with a thickness H; then consolidation settlement can be expressed by:

$$S = \sum_{i}^{n} \frac{\Delta e_i}{1 + e_{0i}} \Delta H_i \qquad (6.8)$$

Using Equation (6.4),

$$S = \sum_{i}^{n} \frac{C_{ci}}{1 + e_{0i}} \Delta H_i \log_{10}(1 + \frac{\sigma_c{'}}{\sigma_{vi}{'}}) \qquad (6.9)$$

There are approximate methods to evaluate C_c value using initial void ratio and liquid limit (Fukue and Mulligan, 2009; Skempton and Jones, 1944). Fukue and Okusa (1987) showed that C_c can be expressed as a function of the initial void ratio, as:

$$C_{ci} = 0.54(e_{oi} - e_{min}) \qquad (6.10)$$

From a practical point of view, e_{min} can be assumed as 0.5 (Fukue and Okusa, 1987).

6.2.2 ROUGH ESTIMATE OF CAP THICKNESS FOR ADVECTION

The total volume of the drained water from sediment can be expressed by:

$$V_d = S\,A \tag{6.11}$$

where A is the area of the cap. Therefore, the penetration of pore water into the cap is given by:

$$p_w = S\,/\,n_c \tag{6.12}$$

where p_w is the penetration distance, and n_c is the porosity of capping materials. Considering the fingering effects for seepage, disturbance effects due to benthos, and drifting of capping materials, the following expression may be used to obtain the cap thickness in which the pore water cannot be released into the water column by advection (Figure 6.3).

$$H_c \geq p_w \tag{6.13}$$

where H_c is the thickness of the cap.

Figure 6.4 shows an example of void ratio profiles for lake sediments. The void ratio was calculated from the water content and the density of particles. The effective pressure was also calculated from the density of particles and the void ratio. The sediment profile was divided into seven parts. The contaminated layer was from the surface to a depth of 46.5 cm. The calculation of settlement S is 23 cm for the top 60 cm of sediment under a σ_c' of 1764 kPa, where the thickness of the sand cap was 30 cm, as presented in Table 6.1. An ordinary sand cap has a unit submerged weight of around 600 to 800 kPa, and may yield a large settlement in the case of soft fine sediments. Therefore, as shown in Table 6.1, p_w values calculated assuming a cap porosity of 0.4 are 58 and 73 cm for 30 and 60 cm cap thickness, respectively. Then, the pore water drained from the sediment can travel through the cap materials (i.e., $p_w > H_c$). This indicates that, if there is no adsorption in the cap layer, the

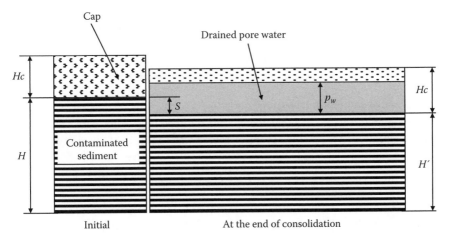

FIGURE 6.3   Illustration of advection flow due to consolidation under the load of a cap.

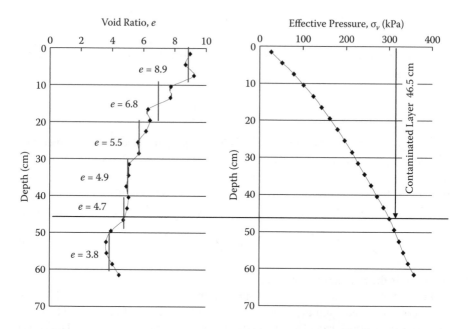

FIGURE 6.4 Measured void ratio profile and calculated effective stress for a soft lake sediment.

contaminants will be released into the water over a long period of time. This may be more dangerous than in the case of no cap, because the advective flow is very low without consolidation due to capping.

Some researchers showed that a good cap thickness was approximately 50 cm. However, it may depend on the consolidation characteristics of the sediments, the bulk density of the sand cap, and the type and concentration of the contaminants. Some contaminants are strongly adsorbed to sediment particles (clay particles), and they are not easily released. On the other hand, the degradation of organic matter releases contaminants into the pore water. Furthermore, organic matter contains high amounts of nutrients. Thus, the pore water can be squeezed through the sand cap. If the level of the released substances is predicted to be too high, other techniques, such as an adsorbent or active mat, should be selected.

6.2.2.1 Contaminant Transport

The contaminant fluxes can be calculated from the beginning of the release until a maximum concentration is obtained in the overlying water using the following equation,

$$J = -D\frac{dC}{dt} \tag{6.14}$$

where J is the flux, dC/dt the concentration gradient as a function of time, and D is the molecular diffusion coefficient. To ensure the short- and long-term effectiveness of the cap, monitoring is required. This will include evaluation of the integrity of the

TABLE 6.1
Calculated Results of Settlement and p_w for Lake Sediments under 30- and 60-cm Sand Caps

30 cm Sand Caps

Depth (cm)	Layer (m)	Void Ratio	C_C	$\frac{C_C}{(1+e_0)}H$	σ_v' (kPa)	σ_c' (30 cm) (kPa)	$\log\left(\frac{\Delta\sigma_c'}{\sigma_v'}\right)$	S (m)	p_w (m)
10	0.1	8.9	4.536	0.05	60	1764	1.483	0.07	
20	0.1	6.8	3.402	0.04	125	1764	1.179	0.05	
30	0.1	5.5	2.7	0.04	195	1764	1.002	0.04	
40	0.1	4.9	2.376	0.04	240	1764	0.922	0.04	
50	0.1	4.7	2.268	0.04	280	1764	0.863	0.03	
60	0.1	3.8	1.782	0.04	320	1764	0.814	0.03	
Total								0.23	0.58
									$n = 0.4$

60 cm Sand Caps

Depth (cm)	Layer (m)	Void Ratio	C_C	$\frac{C_C}{(1+e_0)}H$	σ_v' (kPa)	σ_c' (60 cm) (kPa)	$\log\left(\frac{\Delta\sigma_c'}{\sigma_v'}\right)$	S (m)	p_w (m)
10	0.1	8.9	4.536	0.05	60	3528	1.777	0.08	
20	0.1	6.8	3.402	0.04	125	3528	1.466	0.06	
30	0.1	5.5	2.7	0.04	195	3528	1.281	0.05	
40	0.1	4.9	2.376	0.04	240	3528	1.196	0.05	
50	0.1	4.7	2.268	0.04	280	3528	1.134	0.05	
60	0.1	3.8	1.782	0.04	320	3528	1.080	0.04	
Total								0.29	0.73
									$n = 0.4$

cap through monitoring of the material quality, the thickness of the cap material, and erosion of the cap through resuspension or displacement. Verification must be made particularly after storm events that could have a severe impact on the structure of the cap. It must also be assured that the contaminants are physically isolated by a bathymetric survey, that there is no evidence of chemical diffusion or transport, and that benthic biological recovery is occurring. In other words, the main objectives of the cap must be ensured and determined through monitoring. These are to reduce contaminant fluxes through the cap and to avoid surface sediment and water recontamination. Recovery of the benthic community is another aspect that must be ensured and evaluated during monitoring. Full recovery has been noted at large sites five to seven years after capping (USEPA, 2005).

6.2.3 Active Capping

Other active capping approaches have been used to enhance the reduction of contaminant toxicity and mobility. Reactive caps have shown promise. Organoclays, activated carbon, and apatite have been evaluated (Förstner and Apitz, 2007). Reactive mats with two geotextiles and reactive materials to bind contaminants are also being developed.

A project was carried out by the EPA at the Anacostia River in Maryland (www. ert2.org/sedimentremedy). The sediments contain heavy metals, polycyclic chlorinated biphenyls (PCBs), hydrocarbons, and chlordane. Nearby military and industrial activity were the likely sources of the contaminants. Three active materials, Aquablock™, coke breeze, and apatite, were evaluated as capping materials. Physical, chemical, and biological monitoring was used to evaluate the performance of the caps. Aquablock was effective in terms of physical stability, chemical migration, hydraulic seepage prevention, and effect on flora/fauna such as worms over a 30-month period.

6.3 REHABILITATION OF THE COASTAL MARINE ENVIRONMENT

Because of anthropogenic activities, the coastal marine environment has changed. Many sand beaches have been lost by the reclamation from sea and erosion. Many coastlines have eroded due to a lower supply of sand from river mouths. Since sand beaches have a natural purifying action, the decrease in the number of sand beaches has influenced the quality of seawater.

The excess discharge of wastes into the environment has impacted the coastal marine environment. First of all, nutrients discharged through anthropologic activities have caused eutrophication. Accumulated nutrients in enclosed and semienclosed water areas have led to the formation of red and blue tides which have killed fish.

Hazardous materials have also been discharged into the environment. They have been transported in the form of solutes, adsorbents, or solids (precipitates) into water areas, such as rivers, lakes, ponds, and seas. These substances are potentially taken up by the organisms at the starting point of the coastal marine environment. Subsequently most sea animals become contaminated through the bioaccumulation within the food chain (Bright et al., 1995; de Mora et al., 2004; Hayter, 2006; McLachlan et al., 2001). In this section, rehabilitation techniques for sediments are introduced.

6.3.1 EUTROPHICATION

Eutrophication has been considered to be a phenomenon due to water conditions only. However, it was found that the quality of sediments strongly influences the eutrophication of surface water. This is because sediments contain nutrients which can be released into the overlying water.

In general, nutrients can be released when the organic matter in sediments is degraded. The biodegradation of organic matter causes the decrease in dissolved oxygen (DO), because microorganisms consume DO during their activity under aerobic conditions. The decrease in DO leads to anaerobic conditions, and incomplete degradation takes place. As a result, the production of sulfides, such as hydrogen sulfide, occurs.

Aeration can be a measure to increase DO, but it may stimulate bacteria to degrade organic matter under aerobic conditions. As a result, nutrients will be released into the water again, and eutrophication will continue to occur. Since it is evident that the organic-rich sediment is problematic, it is better to remove some amount of organic matter with nutrients from the water area. This can be achieved by resuspending the sediment particles and filtering them. The idea and experimental results are introduced in a later section.

Another way to control eutrophication may be by sand capping the sediment. However, the reduction of the release of nutrients from the sediment may be only for a short time, because of the time lag for advection and diffusion of nutrients from the sediment. Some data show that the release of nutrients increases after a lag period. In addition, organic matter will biodegrade under anaerobic conditions. This can then lead to the production of gases such as methane, carbon dioxide, and hydrogen sulfide. These gases can accumulate and then diffuse or be transported by advection through the cap material. This can facilitate transport of contaminants though contaminant solubilization or by providing channels through the cap. As long-term measures include capping methods, materials other than sand can be used as described later.

6.3.2 CONTAMINATION

As described earlier in the book, many types of toxic and hazardous substances have been discharged into the environment. The substances have accumulated into the sediments and can be released into water as well as nutrients.

The toxic and hazardous substances are adsorbed, retained, or precipitated. Metals are adsorbed onto the inorganic or organic particles, or precipitated as hydroxides or oxides. Aluminum and magnesium are constituents of mineral particles. Polycyclic aromatic hydrocarbons (PAHs) and volatile organic compounds (VOCs) are also retained in the sediments, in addition to TBT and triphenyl tin (TPT). In general, these substances are concentrated more in the smaller particles and organic matter. The measures for remediation of sediments can be similar to those used for nutrient problems (i.e., removal of organic matter by resuspension, capping, or dredging).

6.3.3   DISTRIBUTION OF CONTAMINATED PARTICLES

Smaller particles have a higher specific surface area than coarse particles. Therefore, the concentrations of retained substances on sediments are influenced by the particle size, as shown in Figure 6.5 and Figure 6.6. The horizontal axis in these figures is taken as content finer than 0.075 mm instead of the particle size. Fukue et al. (2006) showed that particles of grain size diameter at which 10% dry weight are finer ($<D_{10}$) have more than 80% of the specific surface area for the total sediments (Fukue et al., 2006). This suggests that the removal target should be smaller particles (e.g., less than D_{10}). Therefore, techniques for this are needed, because there are many advantages expected as follows:

1. The volume of contaminated sediments can be reduced up to 1/10.
2. Dredging and treatment of dredged materials are not required.
3. Disposal sites for removed materials are minimal.

Dredging is one of the most common sediment remedial actions, but an assessment of case studies showed that remedial dredging often suffers from technical limitations including incomplete removal, unfavorable site conditions, sediment resuspen-

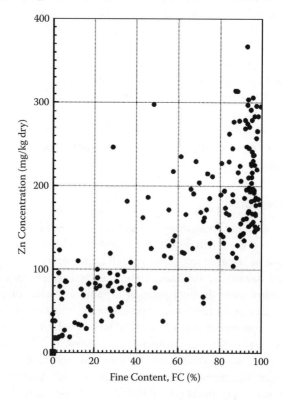

FIGURE 6.5   Zn concentration versus the fine fraction content for various sediments.

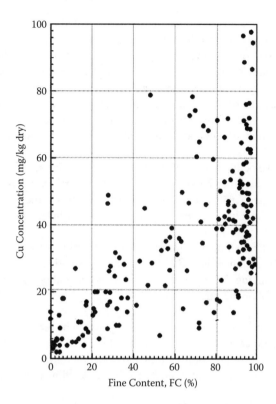

FIGURE 6.6   Cu concentration versus the fine fraction content for various sediments.

sion, and sediment disposal (NRC, 2007). Therefore, new technologies to manage contaminated sediments and new approaches to assess treatment success are needed.

6.3.4   Resuspension Method for Removal of Contaminated Sediment Particles

Resuspension of sediment particles can easily occur naturally or artificially. In nature, resuspension of sediment particles occurs by wave action and currents. Artificial resuspension of sediments occurs during fishing with dragnets, marine construction, propeller wash, dredging, etc. Under this condition, suspended particles are transported by currents and diffusion. If the particles are contaminated, the contamination will spread.

Resuspended particles have a specific gravity greater than one and therefore can settle again. The dispersion height of resuspension and the subsequent settling of particles primarily depend on the grain size and the specific gravity of particles. The maximum dispersion height is greater for smaller or lighter particles. Since organic particles have a lower specific gravity, the resuspension height is larger. Because the smaller-grain-size particles have lower specific gravities and settling velocities, they remain as suspended solids longer.

Organic matter in sediments is the main cause of pollution in water areas, because they retain high levels of hazardous materials, such as heavy metals, viruses, bacteria, etc. In addition, organic matter degradation leads to the anaerobic condition of the sediments. Many studies have shown that organic matter can retain hazardous materials and can be used as an amendment. However, it must be recognized that organic matter can be degraded, and retained materials can be released. Therefore, the removal of organic matter from the bottom of the water column can be one of the ways to remediate the contaminated sediments and to maintain the quality of surface water.

The settling velocity of organic particles is low, because of the smaller specific gravity. On the other hand, relatively large mineral particles can settle fast. Therefore, the organic matter remains as suspended solids in the water column after the inorganic particles have settled. The organic matter can be pumped and removed by filtration.

The resuspension can be achieved mechanically using a stir, water jet, or other techniques. If the sediment is under the anaerobic condition, the water jet with air is effective in changing the sediment to aerobic. For actual performance, the design will include the following information:

1. Total amounts of organic matter (ignition loss) to be removed
2. Contamination degree of the organic matter
3. Filtration method, size of filter unit and selection of filter
4. Treatment and disposal of filtered materials, including reutilization

6.3.5 Technology for Sediment Remediation by Resuspension

The resuspension of sediment particles can be achieved by disturbance of the bottom using physical devices. The use of water jets with air can be the most easy and simple technique. The air injected is for aeration of the sediments. This is important when the sediments are under anaerobic conditions. The mechanical disturbance using the rotation of blades is also a useful way of resuspending the sediment particles, but there are some difficulties if there are buried materials in the surface sediment layer.

The depth of excavation and diffusion of resuspended particles mainly depends on the intensity of the water jet and the grain size distribution. The specific gravity of particles is also an important factor. Smaller and lighter particles are resuspended more. Since the plume of resuspended solids may spread and be transported from site to site, appropriate measures should be used. Silt fences may be useful for shallow depths. If the water depth is high, a cell system or curtain wall should be used. The inlet of the pumping can be adjusted by the SS. If shallow pumping is desired, the intensity of the water jet should be higher. If the amount of pumping SS is too high, the intensity of the water jet should be lowered or stopped.

When other mechanical devices are used for resuspending particles, the situation is similar. Pumping of SS should be controlled by adjusting the elevation of the inlet of the pumping tube, or the intensity of mechanical disturbance. Coarser particles are more resistant to the applied force, and their movement can be limited.

Therefore, by raising the inlet of the pumping tube, smaller particles can be removed. If the specific gravity of particles is smaller, they can travel further away.

The pumped SS includes a large amount of water. Therefore, dewatering is needed. Filtration is the conventional method to remove SS from water. Many kinds of filters, such as particulate media (e.g., sand), geotextiles, etc. are available. Sedimentation for removal of the SS may take longer if there are small or light particles.

6.3.6 DESIGN OF A FILTER UNIT

When filtration is used, the volume of the filtration can be estimated from Darcy's law:

$$\Delta q = kAi\Delta t \quad \text{(in m}^3) \tag{6.15}$$

where
k = coefficient of permeability (m/s)
A = area of filter (m^2)
i = hydraulic gradient
Δt = time
The filtered mass of solids, Δm_i, is

$$\Delta m_i = \alpha \Delta q SS_i / 1000 \quad \text{(in kg)} \tag{6.16}$$

where α is the filtration rate, and SS_i is the suspended solids (mg/L). After Δt, SS in the water system can be given by:

$$SS_{i+1} = \frac{SS_i\{V_i - \alpha \Delta q\}}{V_i - \Delta m_i / \rho_s} \quad \text{(in mg/L)} \tag{6.17}$$

where V_i is total volume of water, and ρ_s is the specific gravity of SS (Figure 6.7).

The calculation result is presented in Figure 6.8. However, clogging of the filter is not taken into account. The clogging properties of the filter depend on many factors, such as the type of the filter, including the pore size, size of SS, initial SS value, filtration rate, etc. Therefore, the average permeability is used. By determining the effects of the variables presented above, a proper filter unit can be designed (Table 6.2). For example, the size of filter can be estimated by changing k, i, or α.

The resuspension technique can be illustrated as shown in Figure 6.9, where sediments are anaerobic. Hydrogen sulfide is produced, and eutrophication is significant. Therefore, the main objective is aeration and the removal of organic matter in water and surface sediments. Monitoring of sediment and water quality must be performed during the remediation.

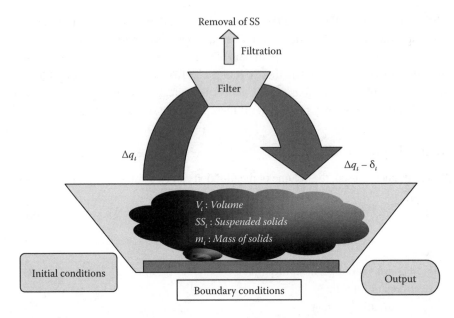

FIGURE 6.7   Conceptual model for removal of resuspended solids.

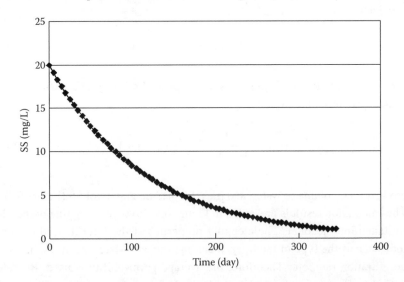

FIGURE 6.8   Example of a calculation for removal of suspended solids.

6.4   CHEMICAL REMEDIATION TECHNOLOGIES

One of the strategies for in situ remediation techniques for sediments is to increase the stabilization of metals on sediment particles (e.g., immobilization). The remediation cost for the stabilization is relatively low. Capping as presented earlier is another method, though it may not impede the release of contaminants from the sediments.

TABLE 6.2

Example of the Parameters for Filtration Design Calculation

Parameter	Value
Characterization of the Water Area	
Area of water system (S)	100 m × 100 m
Average depth (D)	2 m
Water flow in	0
Water flow out	0
Water Quality (Initial)	
SS	20 mg/L
T-P	(not specified)
T-N	(not specified)
COD	(not specified)
Filter Properties	
Average coefficient of permeability (k)	0.00007 m/s
Hydraulic gradient (i)	2
Filter area (A)	200 m^2
Filtration rate (α)	0.7

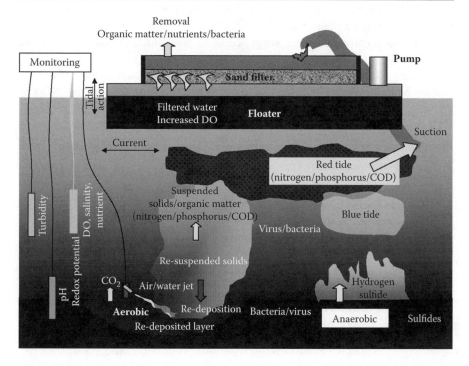

FIGURE 6.9 Illustration of remediation of contaminated sediments by aeration and resuspension.

Stabilization is an approach requiring a simple mixture of amendments. Compared with ex situ remediation techniques, this technique is simpler and lower in cost. However, it may be a major disadvantage that the contaminants still remain in the sediments and thus could possibly leach over time.

Immobilization due to amendment addition is by adsorption to mineral surfaces, formation of stable complexes with organic ligands, surface precipitation, or ion exchange. It will decrease the rate of metal ion leaching and their bioavailability by inducing various sorption processes. Precipitation as oxides or hydroxides and coprecipitation can also contribute to reducing contaminant mobility. These processes are influenced by many factors: pH, redox potential, type of contaminants, cation exchange capacity, type of sediments (type of constituents), etc. The processes are usually combined, because of the varied nature of sediments (Yong et al., 2006). Many kinds of amendments are being used. Activated carbon is one of the most popular amendments (McDonough et al. 2008; Tomaszewski et al., 2008). Kumpiene et al. (2008) reviewed amendments for heavy metals, such as As, Cr, Cu, Pb, and Zn. They evaluated and summarized the effectiveness of trace element immobilization in terms of the type of amendments as shown in Table 6.3.

Sepiolite appears to be an effective amendment to stabilize soil polluted with cadmium and/or zinc. This mineral is able to reduce the soluble cadmium and zinc concentrations of a highly polluted mining soil by 95% and their extractable (soluble exchangeable) concentrations by approximately 84 and 99%, respectively, when a sepiolite dose of 4% was applied to soil (Álvarez-Ayuso and García-Sánchez, 2003). For organic contaminants, such as dichlorodiphenyltrichloroethane (DDT) and PAH, activated carbon has been suggested as an amendment (Brandli et al., 2008; Tomaszewski et al., 2008).

Gardner (2002) showed that the addition of microscale zero-valent iron (ZVI) can dechlorinate PCBs in contaminated sediments. Up to 84% of the PCBs in

TABLE 6.3
Summary of the Effectiveness of Trace Element Immobilization in Terms of Amendment Type

Amendment	As	Cu	Cr	Zn	Pb
Phosphorus materials	−	+		+	++
Organic matter	+/−	+/−	++	+/−	+/−
Clays	+	+		++	+
Alkaline materials	−	+	−	++	+/−
Fe oxides	++	+/−	++	+	+
Mn oxides	++		−		

Source: Adapted from Kumpiene et al., 2008.

Note: (++) = very good, (+) = satisfactory to good, (+/−) = varying results showing weak improvement or both positive and negative effects regarding the element mobility, (−) = should be avoided due to obvious negative effects regarding leaching, () = not found in the reviewed literature.

Houstatonic River sediments were dechlorinated compared to 56% in New Bedford Harbor sediments. PCBs were less strongly sorbed onto the former sediments, and thus the higher reduction. Nanoscale ZVI did not work as well. Lowry et al. (2004) developed an active sediment cap to degrade and sequester the contaminants. ZVI was incorporated into the cap. He found that the microscale ZVI did not react, while the nanoscale ZVI did dechlorinate some PCBs with half-lives of 40 days to 77 years. This contrasts strongly with the results of Gardner (2002). The nanoscale ZVI has a high cost and is not stable for the remediation of strongly sorbed PCBs. However, nanoscale ZVI can be found in soils and sediments which may stimulate anaerobic dechlorination. ZVI is environmentally friendly, because the final product is Fe(III), and it can be easily injected. Native iron-reducing bacteria may help sustain the process. Understanding the fate and transport of nanoscale ZVI and the pathways of dechlorination are needed prior to use.

Chemical oxidants such as potassium or sodium permanganate, hydrogen peroxide, and ozone in various combinations have been used for destruction of organic contaminants. Extraction of organic contaminants can be achieved by injection of steam, cosolvents, or surfactants. In situ chemical oxidation (INCO) was performed at pilot scale at the Upper Main Harbor in Frankfurt, Germany (Thomas et al., 2008). The fine-grained sediment (80% clay and silt) was contaminated with 19 g/kg of total petroleum hydrocarbons (TPH). A 1% solution was injected through 10 screens at a rate of 5 g/kg. Every three hours, an amount of 0.5 L of solution was injected. Some heavy metals were detected in the pore water, posing a risk to the harbor water. Degradation of the organic pollutants was not successful despite local effective peroxide concentrations of 20 g/kg.

More recently, additives to encourage degradation or sequestration of contaminants have been proposed as cap material. Geomembrane materials may be used beneath a cap in soft sediments to aid in the support of the cap and stones, or other large material may be employed as armoring on top of the cap to reduce cap resuspension and erosion. Surficial cap layers may also be designed to improve habitat areas values of the substrate (Reible, 2005). Jacobs and Förstner (1999) suggested that using zeolite as cap material can enable the barrier system to function in an active manner. A pilot study, however, should be performed before applying this in the field.

6.5 BIOLOGICAL REMEDIATION TECHNOLOGIES

An introduction to biological processes was presented in Chapter 3, which was followed, in Chapter 5, by a discussion of natural recovery processes. Bioremediation involves the use of microorganisms to break down the contaminants or convert them to a less toxic form. Unlike natural biodegradation, means are used to improve the rate of biodegradation through addition of nutrients or aeration or to improve the availability of the contaminants. For evaluation of bioremediation potential, site characterization is required. Some parameters include the hydrological characteristics, contaminants (level and type), microbial activity, and the presence of electron donors and acceptors (Rittmann and McCarty, 2001). Permeable sands or gravels are more easily treated by bioremediation than silty or clayey sediments. Heavy metals can be sequestered or precipitated but not biodegraded. Compounds of low water solubilities or high K_{ow} coefficients are less easily biodegraded.

Various field experiments and trials have been performed, mainly following actual spills, to evaluate the feasibility of treatment in the marine environment. The bioremediation of gasoline and fuel oil is well established. The bioremediation of PAHs, chlorinated aliphatics and aromatics, and PCBs is an emerging technology. The presence of non-aqueous-phase liquids (NAPLs), however, decreases the efficiency of the process. Numerous factors have been shown to affect the biodegradation rates of oil, such as the origin and concentration of oil, the availability of oil-degrading microorganisms, nutrient concentrations, oxygen levels, climatic conditions, and sediment characteristics. The application of fertilizers (to obtain specific oil-to-nitrogen ratios) can stimulate the biodegradation rates of oil in aerobic intertidal sediments such as sand. The effect of seeding with hydrocarbon-degrading bacteria has not been clearly beneficial under natural environmental conditions. Many techniques are available for the treatment of oil spills, but guidelines should be established. On the basis of the available evidence, proposed preliminary operational guidelines for bioremediation on shoreline environments have been proposed (Swannel et al., 1996).

Enhancement of the bioremediation of oil contaminated beach sediments has also been performed (Obbard et al., 2004). Crude palm oil, fatty acids, and nutrients (C:N:P = 100:10:1) were added to stimulate the biodegradation of light crude oil. Fatty acid addition led to the complete degradation of straight chain alkanes and enhanced degradation of branched alkanes such as pristine and phytane. The fatty acids function both as cosubstrates and nonionic surfactants.

Seidel et al. (2004) performed a bioremediation feasibility study at pilot scale for heavy metal contaminated river sediments. The process included conditioning in a sludge bed by reed canary grass followed by aerobic solid-bed leaching of heavy metals by Sulfur-oxidizing bacteria and sulfur supplementation. Within 21 days, Zn, Cd, Mn, Co, and Ni were removed at 61% to 81% compared to Cu (21%). Cr and Pb were not immobilized to any extent. Costs of the leaching were 37 to 110 Euro per t (US50 to US$150 per t) for a 10,000 tonne of sediment treatment plant. Metal removal technologies range from €100 to €400/t ($130 to $532/t). Soil washing is the most cost-effective at €30 to €180 per t, but can only be used for sediment if the sand fraction is larger than 30%. Stabilization is considered as not environmentally acceptable by many authorities while incineration is of high cost (€200 to €1000/t or $270 to $1330/t). Disposal is €20 to €200 per t ($30 to $270/t), but will be restricted soon. Benefits include use of ambient temperature, low consumption of energy generation of waste, and no toxic chemicals.

Vezulli et al. (2004) performed a field trial bioremediation study of organic-rich sediments in a fish farm. Both biofixed microorganisms (bioaugmentation) and oxygen-release compounds (ORC) for biostimulation were evaluated. The ORC use increased mineralization and showed a net carbon loss by respiration but only a small increase in carbon mobilization (<10%). Bioaugmentation, on the other hand, increased mobilization by 23%. Therefore, to stimulate physical removal, bioaugmentation may be considered, while biostimulation should be considered for biological removal.

A strategy for stimulating the dechlorination of PCBs involved addition of $FeSO_4$ (Zwiernik et al., 1998). According to Bedard (2003) the addition of only 5 kg of ferrous sulfate (a safe and inexpensive product) has the potential to treat 1 ton of sediment. Another method is through addition of bromobiphenyl. This product, however, is recalcitrant (Abraham et al., 2002). Bioaugmentation of organisms capable of PCB-degradation has not been successful in Housatonic River sediments. Addition of granular anaerobic sludge from an upflow anaerobic sludge blanket (UASB) reactor was evaluated for PCB-contaminated sediments from the River Basin (Natarajan et al., 2000). Significant reduction in tri- to heptachlorobiphenyls was observed. Bedard (2003) indicated, however, that large volumes of granules would be required (up to 10%), and thus the feasibility of full-scale projects is questionable unless advances can be made in granular technology. More field tests are needed.

Another strategy to enhance bioremediation was performed for the treatment of chlorinated methanes, ethenes, and ethanes in a tidal wetland at the Aberdeen Proving Ground in Maryland (Majcher et al., 2009). A biologically reactive two-layer mat with a permeable reactive organic-based bioaugmented matrix was placed on the sediment to react with and reduce contaminants that were transported by an upflow of groundwater. The lower layer contained ZVI filings and an organic matrix including compost, peat, and sand to enhance abiotic degradation of the chlorinated methanes. The upper layer contained compost, peat, chitin, sand, and a microbial consortium for dechlorination (WBC-2). Mass removal of 98% and 94% was achieved for the chlorinated methanes and ethanes/ethenes, respectively. Matrix permeability, metal sequestration, leaching potential, and microbial reactive were evaluated. No adverse affects were observed, and the design was flexible, durable, and effective. The hydrology of the site was also not affected.

The Army Corps of Engineers performed a field-scale bioremediation test in the Mississippi in 2003 (Tiedje, 2004). A sequential anaerobic/aerobic treatment was tested for mineralization of Arochlors 1242/1248. Reductive dechlorination could not be achieved within 6 months despite addition of a carbon source and PCB-contaminated sediment. The 6-month period was probably too short to observe any progress, PCBs are tightly bound to soils and sediments, and also it is difficult to stimulate the dechlorinating microorganisms.

A field-scale aerobic bioremediation study was conducted by General Electric (GE) in 1991 in the Upper Hudson River sediments. Nutrients and hydrogen peroxide (as a supply of dissolved oxygen) (Bedard, 2003) were supplied. PCB destruction was difficult to determine, most likely due to the strong sorption and limited bioavailability of PCBs. Significant mixing to maintain oxygen levels is necessary, but desorption into the water column must be avoided. The availability of the PCBs needs to be improved. Surfactants can be used to enhance solubility, but may inhibit degradation. Ethanol addition is another possibility. Electrokinetics may also enhance availability.

A pilot in situ bioremediation project by in situ aeration (Thomas et al., 2008) was performed for harbor sediments. Pumps, piping, control devices, and storage tanks were installed on a working floating platform to supply oxygen to enhance bioremediation. A flow of 5 to 10 L/hour of air was applied, enabling preferential air flow paths up to 10 m in radius (but mainly in the range of 2 to 4 m). TPH reduction of 60% to 75% and 75% to 85% PAH reduction in sediment cores was obtained after 12 months.

Another study (Thomas et al., 2008) was performed by airlifting contaminated sediments. A floating bioreactor (URS/WCI patent No. 4416591 C1) was used. The sediments would be stripped of the volatiles, and the floating barriers would avoid resuspension of the contaminated sediments. The objective was to recycle the sediments in place and enhance bioremediation by the bubbling within the suspension before settling. The technique was demonstrated at various sites. An acidic lake filled with brown coal fly ash was treated by lifting the ashes with an air/CO_2 mixture to allow the formation of a sustainable calcite buffer. The resuspension was then allowed to settle. This was more viable and sustainable than adding chalk milk to raise the pH of the entire lake (Preuss et al., 2008).

Significantly more experience for in situ bioremediation has been obtained for soil/groundwater systems than for sediments (Fantroussi et al., 2006). Strategies for air sparging or soil vapor extraction (bioventing) have been used to enhance volatilization and bioremediation of contaminants in unsaturated zones. However, sediments are fully saturated, and thus different strategies such as applying water circulation with stimulating agents could be adapted.

Bioaugmentation where specific microorganisms are added has potential but has many difficulties (Fantroussi and Agathos, 2005). It has been demonstrated mainly in laboratory studies. Preadapted pure strains or consortia or genetically engineered cultures may be added. There have been many exaggerations by commercial companies and many failures. *Desulfomonile tiejei* was fed into a 500-L pilot scale reactor to dechlorinate 3-chlorobenzoate (3-CB). Acetate and formate were added as cosubstrates. Polymerase chain reaction (PCR) technology was used to detect 3-CB dechlorination activity.

In another study using a 680-L sandbox, in situ remediation of a trichloroethylene-(TCE) and nickel-contaminated sediment was evaluated by adding methanol and lactate (El Mamouni et al., 2002). Sulfate addition was able to lead to NiS formation by sulfate-reducing bacteria without interfering with TCE transformation to ethane via *cis*-dichloroethene (DCE) and vinyl chloride (VC). Using 16s rDNA, the presence of dechlorinating bacteria of the *Dehalococcoides* group could not be detected, thus indicating the presence of other bacterial groups (Drzyzga et al., 2002).

6.6   CREATION OF SEAWEED SWARDS

Eelgrass (*Zostera marina*) is a water plant with long grasslike leaves. Figure 6.10 shows a dense sward of eelgrass. There are many different species of eelgrass. Shallow intertidal-water eelgrass has shorter and narrower leaves, whereas deeper subtidal-water eelgrass has longer and wider leaves. They tend to grow in tidal creeks, sandy bays, estuaries, and on silty-sandy sediments and are a vital part of the food web chain for the coastal marine ecosystem. In dense swards of eelgrass, silt and clay particles tend to be deposited with organic matter. Decomposition of organic matter will render the seabed anaerobic, and the color of the sediments will become black because of the effect of sulfide.

FIGURE 6.10 Dense sward of eelgrass (*Zostera marina* L.) in a coastal region.

The eelgrass family is one of the few flowering plants that lives in salt water, and the long grass blades are home to various kinds of small marine plants and animals. They are the breeding ground and habitat for all kinds of marine animals including crabs, scallops, and other kinds of shellfish. They not only serve to foster and stabilize the benthic habitats, but they also have the potential for phytoremediation. Eelgrass can absorb trace metals and organotins (Brix and Lyngby, 1982; Fransois et al., 1989).

Table 6.4 gives a comparison of concentrations of various heavy metals in sediments with and without eelgrass. The sediments, which were taken from a small eelgrass sward at an estuary of Kasaoka Bay in Seto Inland Sea, Japan, consisted of a number of small communities—with bare parts between the communities. The total area of the eelgrass sward was 1491 m². The sampling points from A to H were located in the bare parts and communities. The results show that the sediments with eelgrass contain a lesser amount of heavy metals—most likely attributed to heavy metal absorption (uptake) by the eelgrass. Since eelgrass grows from spring to summer, and their dead leaves drift to the sea surface at the end of their growing season, collection of the dead leaves can be simply implemented. This means that, if eelgrass is used for phytoremediation, the absorbed heavy metals can be harvested with the dead seagrass leaves.

Reclamation and other industrial nearshore industrial activities can have a negative impact on the coastal habitat and particularly on the seagrass beds that form the seaweed fields. Reduction of seaweed fields not only decreases the habitat of marine living things, but will also result in a marked decrease in the haul of inshore fish. For example, in Japan, approximately 6000 ha of seaweed field have disappeared since 1978, and about one third of this was due to the impact of reclamation projects.

TABLE 6.4

Comparison of Heavy Metal Concentrations in Sediments with and without Eelgrass (Seto Inland Sea, Japan)

Location	Type of Sample	Heavy Metal Content (mg/kg)		
		Cu	Pb	Zn
A	with eelgrass	11	20	76
B	without eelgrass	13	10	83
C	with eelgrass	11	25	69
D	without eelgrass	14	130	90
E	with eelgrass	15	17	77
F	with eelgrass	16	18	82
G	without eelgrass	27	17	110
H	with eel grass	17	18	83
	average with eelgrass	14	19.6	77.4
	average without eelgrass	18	52.3	94.3

Source: Adapted from Yong et al., 2006.

Recently, a 60-week phytoremediation feasibility test was performed with eelgrass for treatment of PAH- and PCB-contaminated sediments (Huesemann et al., 2009). PAH levels decreased by 73% compared to 25% in the unplanted control area. PCB removal was slightly less (60%). However, no removal was seen in the unplanted area. The mechanism of remediation was biodegradation in the root area. There appeared to be minimal limitation for mass transfer and bioavailability because PAHs and PCBs were found in the roots and shoots. However, overall, only 0.5% of the total amount of PAHs and PCBs were translocated into the plant. Therefore, the main mechanism seemed to be the stimulation of biodegradation by the presence of the eelgrass due to the release of plant enzymes or oxygen. More research is needed to study the mechanisms of removal.

6.7 CASE STUDIES OF REMEDIATION

6.7.1 CONTAMINATED SEDIMENT CAPPING PROJECTS

According to the list published by Hazardous Substance Research Centers (HSRC) (http://www.sediments.org/capsummary.pdf), one of the oldest contaminated sediment capping projects was performed in New England. It consisted of 52 small projects which started in 1980. An approximate 50-cm silt cap was used to cover metal- and PAH-contaminated sediments. In the same year, a capping project was initiated at Hiroshima Bay, Japan. To combat Minamata disease, a record 2.8-m-thick cap was achieved. This is probably one of the maximum thicknesses used for capping.

The widest capping site was East Sha Chan, where there were mud pits contaminated with copper and chromium in Hong Kong. The project was divided into two periods (i.e., from 1992 to 1997 and from 1997). The cap area of the former was about 2.3 million m², and the latter was 2 million m². The cap materials used were clean sand and mud. It can thus be seen from the list that capping has been used for a variety of contaminants, such as heavy metals, PCBs, PAHs, TBT, polychlorinated dibenzofurans (PCDF), DDT, oil and grease, nutrients, etc.

6.7.2 STEEL SLAG

Sulfide is an environmental index for marine sediments, because sulfur reacts with hydrogen or metals under anaerobic conditions. Among the sulfides, hydrogen sulfide is one of the most toxic materials. Therefore, one of the urgent problems in canals, ports and harbors is to suppress the occurrence of hydrogen sulfide from the sediments. In this section, the effects of capping using steel slag on suppression of sulfide was examined.

A mesh container of 260 × 190 × 235 mm was used to measure the concentration of sulfide, as a small capping model, as shown in Figure 6.11(b). Steel slag particles with a diameter between 30 and 40 mm were put in the container and placed on the sea bottom in Orido Bay (Shimizu Port, Japan). A comparison with other materials, granite and concrete pebbles, as capping materials was performed. On the other hand, containers with a dimension of 200 mm in diameter and 210 mm high were used to investigate the diversity of the benthos. A similar container was used for granite pebbles, which were up to a height of 100 mm, as shown in Figure 6.11(a).

After two months, the water samples were obtained from the bottom, middle, and top of the container. The concentration of volatile sulfide with acid was measured using a test kit. The concentrations of sulfide and pH are shown in Figure 6.12. It is noted that sulfide was not detected in the water samples from the steel slag, while the concentrations varied between 0.2 and 0.25 mg/L on the samples from the granite and concrete pebbles, respectively. These results may indicate that the slag can suppress the release of sulfide from the sediments.

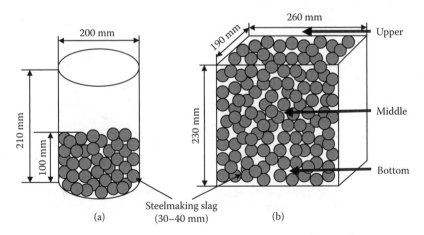

FIGURE 6.11 Model of a cap using the steel slag, granite, and concrete pebbles.

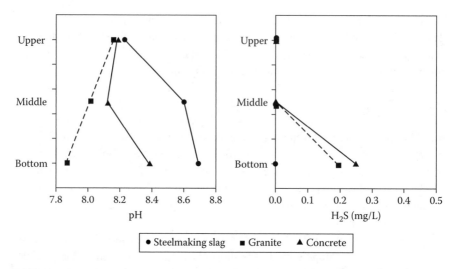

FIGURE 6.12   pH and H₂S changes in the different capping materials, steelmaking slag, granite, and concrete pebbles.

After one month, the containers were recovered, and the number of benthos was determined. Figure 6.13 shows the number and wet weight of benthos from the three containers with different materials. It can thus be concluded that the steel slag showed a beneficial effect on the diversity of benthos.

Steel slag may suppress the release of sulfide from sediments. This may indicate that steel slag can be used as capping materials for anaerobic sediments. However, it should be verified whether any pollutants are released from the slag itself under any condition.

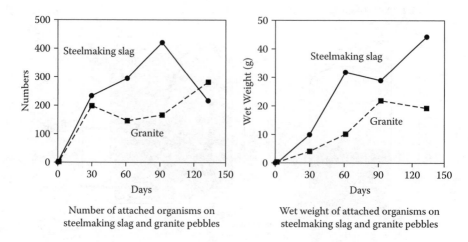

FIGURE 6.13   Increased numbers of benthos in the caps using steel slag and granite pebbles, respectively, each month.

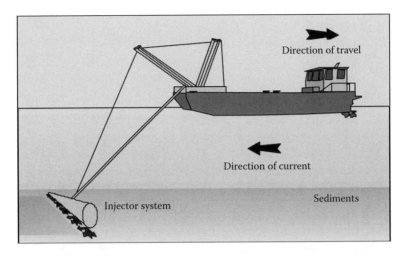

FIGURE 6.14   LIMNOFIX process for in situ bioremediation (Guo and Murphy, 2008).

6.7.3 BIOREMEDIATION

Various in situ bioremediation approaches have also been evaluated. An example is the LIMNOFIX In Situ Treatment Technology (LIST) that was developed for the in situ remediation of contaminated sediment in fresh and marine water environments (Guo and Murphy, 2008). The system consists of a chemical delivery system, mounting platform for equipment (Figure 6.14) and chemical storage, and a chemical formation including oxidants (calcium nitrate), binders, flocculants, and/or other amendments. The oxidants can reduce odor, nutrient release, and sulfide toxicity. Approximately 5,000 m^3 of sediments contaminated with PAH and TPH were treated at Hamilton Harbor, Ontario. Full-scale treatment was performed in the United States in 1998 at a coal tar-contaminated intertidal zone (90% reduction of PAHs and 50% reduction of TPH) and at the Shing Mun River in Hong Kong (90% to 99.9% reduction of sulfide). The injection of calcium nitrate facilitates aerobic bioremediation. The bacteria convert nitrate to nitrogen gas and the organics to carbon dioxide and water. Over 230,000 m^3 of riverbed were bioremediated. Acid volatile sulfide (AVS) was reduced from 2100 to less than 5 mg/kg. The process was effective to 20 cm of sediment depth. Results of the various studies are shown in Table 6.5.

An in situ bioremediation method was demonstrated using a hydraulic dredge at a basin with contaminated sediments (Paquin, 1994). Effluents were sent to the basin from a petrochemical plant over a period of 14 years. The setup of the remediation is shown in Figure 6.15. More than 1350 m^3 of sediments were contaminated with 1.4% mineral oil and grease. A hydraulic dredge was used to mix the sediments with alkali, nutrients, biosurfactant, and enzymes in the lagoon of 375 m^2. The purpose of adding the biosurfactants and enzymes was to stimulate the indigenous microbial population by solubilization and chemical breakdown of the contaminants. Oxygenation was also added via a portable compressor. For the first four weeks, the rate of biodegradation was 340 mg/kg-day, but this decreased to less than 50 mg/kg-day. However, after 10 weeks the cleanup objective of 0.1% oil and grease level was

TABLE 6.5

Summary of the Results for the LIMNOFIX Process

Site (year)	Scale	Contaminants	Concentration (mg/kg)		Reduction %
			Initial	Final	
Aluminum Co. (1996)	Bench	TPH	27.7	5.4	81
Aluminum Co. (1996)	Bench	PAHs	30,800	16,200	47
Refinery Oil Lagoon Sludge (1) (1994)	Bench	PAHs	5,300	2,200	59
Refinery Oil Lagoon Sludge (1) (1994)	Bench	TPHs	350,000	140,000	60
Refinery Oil Pond Sediment (2000)	Bench	Sulfide	12,700	350	97
Refinery Oil Pond Sediment (2001)	Bench	Sulfide	13.7–35,300	1.46–3,400	91–99
Hamilton Harbour (1992–1994)	Pilot	PAHs	730	260	64
Hamilton Harbour (1992–)	Pilot	TPH	11,800	5,000	57
Hamilton Harbour (1992–1994)	Pilot	BTEX	0.24	0.05	79
Hong Kong (1997–1998)	Pilot	Sulfide	4.630	360	92
St. Mary's River (1991–1993)	Pilot	Sulfide	1,450	290	80
Salem, MA (2) (1998–)	Full	PAHs	115	10	90
Salem, MA (2) (1998–1999)	Full	TPH	400	200	50
Shing Mun River (2001)	Bench	Sulfide	450–2092	21.2–50.2	96–97
Shing Mun River (2002)	Full	Sulfide	150–5800	<5–310	>99.9–90.3

Source: Modified from Guo and Murphy, 2008.

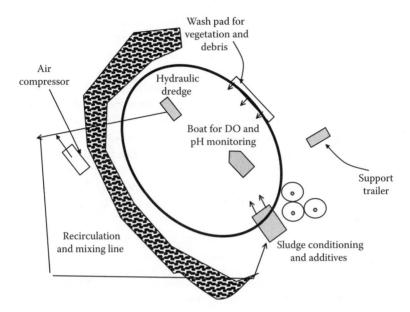

FIGURE 6.15 Arrangement of a biotreatment system developed by Sanexen (Paquin, 1994).

obtained. Overall, the stimulation enhanced the natural processes by approximately 100 times. The cost of the bioremediation was $75/tonne (1994 dollars). The site was decontaminated sufficiently for reuse (criteria B of Environment Canada).

6.8 CONCLUDING REMARKS

In this chapter, in situ remediation techniques for sediments are described. Capping is the most common technique used as an in situ remediation technique. In this technique, the advection and diffusion of contaminants are often discussed in terms of consolidation. At present, many models have been developed to predict the transport of contaminants.

On the other hand, there may be a lack of knowledge in the physicochemical properties of sediments, especially sorption and desorption or release of substances. Sediments have already been washed by water during their settling. This means that the adsorbed substances on the particles cannot be easily released, unless pH or redox potential changes. Therefore, the change or variation has to be taken into account in the analyses.

Capping may promote the consolidation of sediments, which may cause an adverse effect for advection transport of contaminants. The calculation using actual data obtained from a small lake showed that the sand capping will cause a large amount of consolidation of soft sediments, and the drained water can possibly penetrate the cap materials. The condition $H_c > pw$ was rarely obtained.

In situ chemical remediation consists mainly of the addition of amendments. In this approach, the target substances and the corresponding amendment are important.

The literature shows the significant benefits for rehabilitation of sediments, but adverse effects should be examined by using the amendment under a variety of environmental conditions. The major advantage of in situ remediation is that treatment with dredging can be achieved, which may allow remediation to be achieved with minimal disruption of the ecosystem. Costs are usually lower than dredging also. Monitoring is particularly key for verifying that the remediation is taking place and that there is little or no impact over the short and long term.

Although some studies have been performed for in situ bioremediation, the list has not been extensive. Most of the cases are related to the remediation of nearshore sediments after oil spills. Although bioremediation can be cost effective compared to other remediation processes, there are many difficulties, such as bioavailability. Methods are required to enhance nutrient delivery and contaminant accessibility.

REFERENCES

Abraham, W.R., Nogales, B., Golyshin, P.N., Pieper, D.H. and Timmis, K.N. 2002. Polychlorinated bi-phenyl-degrading communities in soils and sediments. *Curr. Opin. Microbiol.* 5(3): 246–253.

Álvarez-Ayuso, E. and García-Sánchez, A. 2003. Sepiolite as a feasible soil additive for the immobilization of cadmium and zinc. *Sci Total Environ.* 305: 1–12.

Arega, F. and Hayter, E. 2008. Coupled consolidation and contaminant transport model for simulating migration of contaminants through the sediment and a cap. *Appl. Math. Modell.* 32: 2413–2428.

Azcue, J.M., Zeman, A.J., Mudroch, A., Rosa, F., and Patterson, T. 1998. Assessment of sediment and porewater after one year of subaqueous capping of contaminated sediments in Hamilton Harbour, Canada. *Water Sci. Technol.* 37(6–7): 323–329.

Bedard, D.L. 2003. Polychlorinated biphenyls in aquatic sediments: Environmental fate and outlook for biological treatment. In M.M. Haggblom and I. Bossert (Eds.), *Dehalogenation: Microbial Processes and Environmental Applications.* Kluwer Press, Boston, pp. 443–465.

Berg, U., Neumann, T., Donnert, D., Nüesch, R., and Stüben, D. 2004. Sediment capping in eutrophic lakes–efficiency of undisturbed calcite barriers to immobilize phosphorus. *Appl. Geochem.* 19: 1759–1771.

Brandli, R.C., Hartnik, T., Henriksen T., and Cornelissen, G. 2008. Sorption of native polyaromatic hydrocarbons (PAH) to black carbon and amended activated carbon in soil. *Chemosphere* 73(11): 1805–1810.

Bright, D.A., Dushenko, W.T., Grundy, S.L., and Reimer, K.J. 1995. Effects of local and distant contaminant sources: Polychlorinated biphenyls and other organochlorines in bottom-dwelling animals from an Arctic estuary. *Sci. Total Environ.* 160/161: 265–283.

Brix, H. and Lyngby, J.E. 1982. The distribution of cadmium, copper, lead, and zinc in eelgrass (*Zostera Marina* L.). *Sci. Total Environ.* 24: 51–63.

de Mora, S., Fowler, S.W., Wyse, E., and Axemard, S. 2004. Distribution of heavy metals in marine bivalves, fish and coastal sediments in the Gulf and Gulf of Oman. *Mar. Poll. Bull.* 49: 410–424.

Drzyzga, O., El Mamouni, R., Agathos, S.N., and Gottschal, J.C. 2002. Dehalogenation of chlorinated ethenes and immobilization of nickel in anaerobic sediment column under sulfidogenic conditions. *Environ. Sci. Technol.* 36: 2630–2635.

El Mamouni, R., Jacquet, R., Gerin, P., and Agathos, S.N. 2002. Influence of electron donors and acceptors on the bioremediation of soil contaminated with trichloroethene and nickel: Laboratory- and pilot-scale study. *Water. Sci. Technol.* 45: 49–54.

Environment Canada and Ministère du Développement durable, de l'Environnement et des Parcs du Québec. 2007. Criteria for the Assessment of Sediment Quality in Quebec and Application Frameworks: Prevention, Dredging and Remediation.

Fantroussi, S.E. and Agathos, S.N. 2005. Is bioaugmentation a feasible strategy for pollutant removal and site remediation? *Curr. Opin. Microbiol.* 8: 268–275.

Fantroussi, S.E., Agathos, S.N., Pieper, D.H., Witzig, R., Camara, B., Gabriel-Jurgens, L., Junca, H., Zanaroli, G., Fava, F., Perez-Jimenez, J.R., Young, L.Y., Hamonts, K., Lookman, R., Maesen, M., Diels, L., Dejonghe, W., Dijk, J., and Springael., S. 2006. Biological assessment and remediation of contaminated sediments, In D. Reible and T. Lanczos (Eds.), *Assessment and Remediation of Contaminated Sediments.* Springer, The Netherlands, pp. 179–238.

Förstner, U., and Apitz, S. 2007. Sediment Remediation: U.S. Focus on Capping and Monitored Natural Recovery. *J. Soil Sed.* 7 (6): 351–358.

Fransois, R., Short, F.T., and Weber, J.H. 1989. Accumulation and persistence of tributyltin in eelgrass (*Zostera Marina* L.) tissue. *Environ. Sci. Technol.* 23(2): 191–196.

Fukue, M. and Mulligan, C.N. 2009. Development of a theoretical approach for prediction of soil compression behaviour. *Soils Foundations* 49(1): 99–114.

Fukue, M. and Okusa, S. 1987. Compression law of soils. *Soils Found.* 27(1): 23–34.

Fukue, M., Okusa, S., and Yoshimoto, N. 1987. General characteristics of upper soil sediments. *Mar. Geotechnol.* 7: 15–36.

Fukue, M., Yanai, M., Sato, Y., Fujikawa, T., Furukawa, Y., and Tani, S. 2006. Background values for evaluation of heavy metal contamination in sediments. *J. Hazard. Mat.* 136: 111–119.

Gardner, K. 2002. In situ treatment of PCBs in marine and freshwater sediments using colloidal zero-valent iron: CICEET progress report for the period 03/01/02 through 09/01/02. Cooperative Institute for Coastal and Estuarine Environmental Technology (CICEET). http://ciceet.unh.edu/progressreports/2002/fall/gardner/index.html.

Guo, J. and Murphy, T. 2008. Limno-fix in situ sediment treatment. In Situ Contaminated Sediments Workshop, Concordia University, Montreal, Quebec, March 10–11.

Hamer, K. and Karius, V. 2005. Tributyltin release from harbour sediments—Modelling the influence of sedimentation, bio-irrigation and diffusion, using data from Bremerhaven. *Mar. Poll. Bull.* 50: 980–992.

Hayter, E.J. 2006. Evaluation of the state-of-art contaminated sediment transport and fate modeling system. EPA/600/R-06/108. U.S. Environmental Protection Agency, National Exposure Research Laboratory, Research Triangle Park, NC.

Hazardous Substance Research Centers (HSRC). 2002. Summary of Contaminated Sediment Capping Projects. http://www.hsrc_ssw.org/capsummary.pdf.

Huesemann, M.H., Hausmann, T.S., Fortman, T.J., Thom, R.M, and Cullinan, V. 2009. In situ phyto remediation of PAH and PCB contaminated marine sediments with eelgrass (*Zostera marina*). *Fifth International Conference on Remediation of Contaminated Sediments*, Feb. 2–5, 2009, Jacksonville, FL.

Jacobs, P.H. and Förstner, U. 1999. Concept of subaqueous capping of contaminated sediments with active barrier systems (ABS) using natural and modified zeolites. *Water Res.* 33(9): 2083–2087.

Kumpiene, J., Lagerkvist, A., and Maurice, C. 2008. Stabilization of As, Cr, Cu, Pb and Zn in soil using amendments—A review. *Waste Manag.* 28: 215–225.

Lowry, G., Murphy, P.J., and Johnson, K.M. 2004. Development and in situ application of sorbent/reagent-amended "active" sediment caps for managing HOC-contaminated sediments. Abstract from the Technology Benchmarking Workshop on Sediment and Floodplain Remediation, March 25–26, 2004, University of Michigan.

Majcher, E., Lorah, M.M., and Graves, D. 2009. Design and performance of an enhanced bioremediation treatment approach at the sediment/surface water interface. *Fifth International Conference on Remediation of Contaminated Sediments*, Feb. 2–5, 2009, Jacksonville, FL.

McDonough, K.M., Fairey, J.L., and Lowry, G. V. 2008. Adsorption of polychlorinated biphenyls to activated carbon: Equilibrium isotherms and a preliminary assessment of the effect of dissolved organic matter and biofilm loadings. *Water Res.* 42: 575–584.

McLachlan, M.C., Haynes, D., and Müller, J.F. 2001. PCDDs in the water/sediment–seagrass–dugong (*Dugong dugon*) food chain on the Great Barrier Reef (Australia). *Environ. Poll.* 113: 129–134.

Mohan, R.K., Brown, M.P., and Barnes, C.R. 2000. Design criteria and theoretical basis for capping contaminated marine sediments, *App. Ocean Res.* 22: 85–93.

Moo-Young, H., Myers, T., Tardy, B., Richard, L., Vanadit-Ellis, W., and Sellasie, K. 2001. Determination of the environmental impact of consolidation induced convective transport through capped sediment, *J. Hazard Mat.* 85(1–2): 53–72.

Natarajan, M.R., Nye, J., Wu, W.M., Wang, H., and Jain, M.K. 2000. Reductive dechlorination of PCB-contaminated Raisin River sediments by anaerobic microbial granules. *Biotechnol. Bioeng.* 55(1): 182–190.

NRC. 2007. *Sediment Dredging at Superfund Megasites: Assessing the Effectiveness.* National Research Council of the National Academies, Committee on Sediment Dredging at Superfund Megasites. The National Academies Press, Washington, DC.

Obbard, J.P., Ng, K.L., and Xu, R. 2004. Bioremediation of petroleum contaminated beach sediments: Use of crude palm oil and fatty acids to enhance biodegradation. *Water Air Soil Poll.* 157: 149–161.

Olsta, J.T., Hornaday, C.J., and Darlington, J.W. 2006. Reactive material options for in situ capping. *Journal of ASTM International* 3(6), Online ISSN: 1546-962X, Published Online: 24 May 2006.

Palermo, M. R. 1998. Design considerations for *in-situ* capping of contaminated sediments. *Water Sci. Technol.* 37(6–7): 315–321.

Paquin, J. 1994. Demonstration of dredging and an additive/aeration system to promote the biodegradation of petrochemical in wastewater sediments. *4th Annual Symposium on Groundwater and Soil Remediation*, Sept. 21–24, 1994, Calgary, Alberta, Canada.

Preuss, V., Koch, T., and Thomas, J. 2008. Contaminated sediment treatment with the floating bioreactor. http://www.fona.de/pdf/forum/2008/abstracts/pV.03_thomas_abstract_forum_2008.pdf.

Reible, D. 2005. In situ sediment remediation through capping: Status and research needs. Hazardous Substance Research Centers, South and Southwest. http://www.hsrc-ssw.org/pdf/cap-bkgd.pdf.

Rittmann, B.E. and McCarty, P.L. 2001. *Environmental Biotreatment: Principles and Applications*, New York, McGraw Hill.

Schauser, I., Hupfer, M., and Brüggemann, R. 2004. SPIEL—A model for phosphorus diagenesis and its application to lake restoration. *Ecol. Modell.* 176: 389–407.

Seidel, H., Lser, C., Zehnsdorf, A., Hoffmann, P., and Schmerold, R. 2004. Bioremediation process for sediments contaminated by heavy metals: Feasibility study on a pilot scale. *Environ. Sci. Technol.* 38(5): 1582–1588.

Skempton, A.W. and Jones, O.T. 1944. Notes on the compressibility of clays. *Quarterly J. Geol. Soc. London* 100(1–4): 119–135.

Swannell, R. P. J., Lee, K. and McDonagh, M. 1996. Field evaluations of marine oil spill bioremediation. *Microbiological Rev.* 60: 342–365.

Terzaghi, K. and Peck, R.B. 1967. *Soil Mechanics in Engineering Practice.* John Wiley & Sons, Inc., New York.

Thomas, J., Beitinger, E., Grosskinsky, H., Koch, T., and Preuss, V. 2008. Innovative in situ treatment options for contaminated sediments. *5th International SEDNET Conference,* May 27–29, Oslo, Norway.

Tiedje, J. 2004. Telephone interview. Michigan State University, June with A. Mikszewski, National Network for Environmental Management Studies Fellow for the preparation of Emerging Technologies for the In situ Remediation of PCB-Contaminated Soils and Sediments: Bioremediation and Nanoscale Zero-Valent Iron, Aug. 2004 for the Environmental Protection Agency.

Tomaszewski, J.E., McLeod, P.B., and Luthy, R.G. 2008. Measuring and modeling reduction of DDT availability to the water column and mussels following activated carbon amendment of contaminated sediment, *Water Res.* 42: 4348–4356.

USEPA. 2005. *Contaminated Sediment Remediation Guidance for Hazardous Waste Sites* EPA540-R-05-012. U.S. Environmental Protection Agency, Office of Solid Waste and Emergency Response, Office of Solid Waste and Emergency Response, Washington, DC. www.clu-in.org.

Vezulli, L., Pruzzo, C., and Fabiano, M. 2004. Response of the bacterial community to in situ bioremediation of organic-rich sediments. *Marine Poll. Bull.* 49: 740–751.

Yong, R.N., Mulligan, C.N., and Fukue, M. 2006. *Geoenvironmental Sustainability.* Taylor and Francis, Boca Raton, FL.

Zwiernik, M.J., Quensen III, J.F., and Boyd, S.A. 1998. $FeSO_4$ amendments stimulate extensive anaerobic PCB dechlorination. *Environ. Sci. Technol.*, 32(21): 3360–3365.

7 Dredging and the Remediation of Dredged Contaminated Sediments

7.1 INTRODUCTION

Dredging is the excavation of materials (sediments) from the bottom of the water column for a number of different purposes, as follows.

1. Navigation: to maintain navigation depths
2. Construction and reclamation: to excavate the bottom for foundations of structures, such as breakwaters, bridges, pipe lines, etc.
3. Purification of surface water: to clean up the bottom of rivers, lakes, ports, and harbors to purify the surface water
4. Environmental: to remove contaminated sediments to protect aquatic life and preserve the safety of seafood
5. Mining: to obtain coarse materials for construction materials

The objective sites for dredging are seas, canals, rivers, ponds, lakes, etc. To maintain navigable waterways, approximately 306 million cubic meters of material are dredged in the United States every year. Of this amount, about 46 million cubic meters are placed in ocean waters at more than 100 Environmental Protect Agency approved sites (http://www.globalsecurity.org/military/systems/ship/dredges.htm). It is also estimated that another 260 million cubic meters are dredged in coastal and inland waters and placed in a variety of locations, including uplands, beach sites, wetlands construction sites, and riverine sandbars, etc. Dredged material in the United States cannot be dumped at sea if it is toxic according to laboratory tests or if, in separate tests, certain chemicals have accumulated in the tissues of exposed organisms (USEPA/USACE, 1991).

Dredging involves removal of the sediment from the waterway with either mechanical buckets or hydraulic pumping. Mechanical dredging removes the sediment at the same water content as that found in situ. This minimizes the amount of water that will need to be processed and treated. The most common mechanical dredges include clamshell, enclosed bucket, and articulated mechanical (Palermo et al., 2004).

Hydraulic dredges remove the sediment as a slurry, and thus the solids content is less than by mechanical dredging. Hydraulic dredges include cutterhead, horizontal auger, plain suction, pneumatic, specialty dredgehead, and diver-assisted types (Palermo et al., 2004). They reduce the potential for sediment resuspension, but may not be able to handle debris and can be less productive. Pneumatic dredges may have significant air entrainment. Diver-assisted can be very precise, but productivity can be low and there is the potential for residuals.

The dredging process itself has the potential to impact the environment. Bray (2007) has listed criteria to consider including:

- Safety of personnel working on the project and near the area
- Accuracy of the contamination area (minimization of dredged material while assuring removal of the contaminated material)
- Generation of suspended sediment
- Mixing of various sediment layers
- Spreading of the contaminated sediment via dilution and transport
- Noise and air pollution during dredging and associated activities

Proper design of the dredging project can minimize the environmental impact. Pilot testing and modeling may be used to predict sediment transport. Some factors include the choice of the type of dredger and operation conditions, the use of mitigating measures, such as the use of silt curtains to prevent the spread of sediments, and the management of the dredged sediments. Long-term monitoring is rarely performed to determine the residual contamination and long-term effects of the dredging.

In the ports of the Severn Estuary in the United Kingdom, it was reported that around 4.5 million tonnes of sediments were dredged in a typical year (Severn Estuary Strategy 1997), whereas in Strangford Lough only 2,000 tonnes were dredged annually (http://www.ukmarinesac.org.uk/activities/ports/ph5.htm).

In Japan, dredged materials of 1.5 to 1.8 million cubic meters from the ports and harbors, except for fishing ports, have been disposed offshore every year. This range is approximately 3% of the total dredged materials per year. Based on the protocol of the London Convention, the Ministry of Environment, Government of Japan, issued a guideline for offshore disposal of dredged materials in 2006.

There are different guidelines for the disposal of dredged material (DelValls et al., 2004). London Convention 1972 (LC) (www.Londonconvention.org), Oslo/Paris Convention (OSPAR) (www.ospar.org), and the Helsinki and Barcelona Conventions have provided the basis for the guidelines. The three conventions suggest the use of different methodologies from physicochemical to biological approaches to the management of different routes of disposal or uses of the dredged material. Most of these conventions propose methods based on a "weight of evidence" (WOE) approach (DelValls et al., 2004). Meegoda (2008) indicated that in New York/New Jersey ports for dredged sediments, 2.3 million tons of sediments can be ocean disposed, 3.1 million tons can be deposited in the ocean with capping, and 1.6 million tons cannot be ocean disposed due to failed criteria for toxicity or bioaccumulation.

At present, different countries have different scenarios and are seeking harmonization in treating dredged materials (Bolam et al., 2006; DelValls et al., 2004; Petrovic and Barceló, 2004; New Delta, http://www.newdelta.org/temp/455321265/

NEW!_Delta_Theme_6_report_6-2_-_07_07_04.pdf). The contents of the Helsinki and Barcelona Convention were reported by Selin and VanDeveer (2002).

7.2 SUSTAINABLE DREDGING STRATEGIES

There may be various approaches to establish sustainable dredging strategies, associated with many factors, such as local differences in regulations. It seems that the varied nature of sediment contamination has created serious and complex problems. However, the London and Helsinki and Barcelona conventions may play an important role in establishing the sustainable development of port and port-related activities and control and management of contaminated sediments in coasts and estuaries. In principle, the convention protocols apply equally to all member states. Nevertheless, different countries implement them in different ways, because of different stages in development and local differences in regulations and cultural factors.

As stated in a report by NEWdelta, at present, a major challenge is to have the various stakeholders working together to manage the estuaries while balancing economic, environmental, and safety aspects. The overall aim is therefore to harmonize the accessibility of ports with the preservation of nature and, at the same time, to improve safety by reducing flood risks and potential for marine accidents (http://www.newdelta.org/temp/455321265/NEW!_Delta_Theme_6_report_6-2_-_07_07_04.pdf).

Sustainable dredging strategies are summarized in Figure 7.1. In the near future, dumping of dredged materials offshore will be prohibited. Japan has decreased the dumping volume of dredged materials and has been preparing for the nondumping stage. As shown in the figure, technologies for detoxification, degradation of toxic organic compounds, and solidification/immobilization of toxic substances are very important. In particular, new cost-effective, simple, and technically effective technologies are desired. Land storage of dredged materials is problematic due to the large amounts that need to be handled and potential leaching of the pollutants that could recontaminate rivers and other water bodies. Combining dredging with other techniques including backfilling, monitored natural recovery (MNR), or capping may reduce dredging requirements and impact. Long-term recovery of the biota must be evaluated at the dredging site.

Strict prohibitions for sea disposal are articulated in the London Convention and Protocols. Dumping or discharging land-based industrial waste into the sea is essentially prohibited—with the burden of responsibility resting on the waste generator to ensure that any waste material entering the sea must be nontoxic and nonhazardous. Since many countries and jurisdictions with restricted land areas do not have sufficient land space for land disposal of waste, controlled and regulated discharge of municipal and industrial wastes into the sea remains as the option of last resort. When such a need arises, waste disposal sites in the sea must be constructed to meet safety and health protection requirements. Isolation of the waste from contact and dispersion into the sea is a prime requirement. In some countries, artificial islands have been constructed for the principal purpose of emplacing secure disposal facilities. These island-based disposal sites must conform to all the regulations that attend

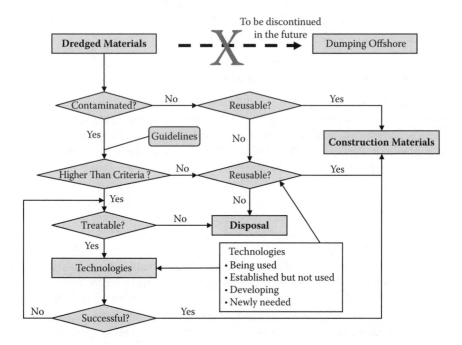

FIGURE 7.1   Sustainable dredging strategies.

land-based disposal sites—with the strict requirement for monitoring and control, to ensure no escape of leachate into the sea.

In some other cases, actual disposal sites have been constructed in the sea using seawalls as containment walls to prevent escape of waste and leachates into the surrounding sea. Typical examples are seen in Tokyo Bay and Osaka Bay. The objectives of the Phoenix project not only focused on the proper and safe disposal–discharge of the wastes in Osaka Bay, but also on preservation of the coastal marine ecosystem and development of the regional area through environmentally acceptable reclamation of shoreline.

7.3   PHYSICAL REMEDIATION TECHNOLOGIES

Two options are available for disposal of dredged contaminated sediments: (a) disposal in a secure landfill and (b) treatment of the contaminated sediments and reuse of the treated sediments. Option (a) is not an option that has many proponents. Treatment of contaminated sediments (option (b)) can be an expensive procedure, especially when the quantities are large. An expedient procedure is to perform gravity separation of the contaminated sediment and to remove the coarse fractions for treatment and reuse as construction material. A useful technique for sediments that do not contain much organic matter is to form larger particles by promoting aggregation of the fines with lime. Contaminated fine fractions can be treated or disposed of in secure settling ponds. These settling ponds are not unlike those obtained in natural resource extraction processes. Techniques for dewatering and hastened sedimentation of the suspended fines that constitute the fine fractions of the sediment have been discussed in Yong et al. (2006). In the case of the fines in sediments, solidification and compression by filter pressing can be used (Yamasaki et al., 1995).

7.3.1 PHYSICAL SEPARATION

Physical separation processes are generally technically simple methods for separation of dredged solids on the basis of size and density and are often used as pretreatments. These processes have been applied for the selection of contaminated fractions and clean coarser particles. Because coarser particles, such as sand and gravel fractions, have less activity on their surfaces, washing is often enough to clean for beneficial use. This is important to reduce the amount of dredged materials to be disposed in confined disposal facilities or in open water. The most contaminated fractions may require further treatment or restricted disposal. The volume of the fine residuals may be minimized using mechanical dewatering techniques (Olin-Estes and Palermo, 2001).

Physical separation processes are used to remove smaller, more contaminated particles. These processes include centrifugation, flocculation, hydrocyclones, screening, and sedimentation. Hydrocyclones can be used for sediments with less than 20% solids to separate coarse or fine-grain fractions. They include hydrocyclones, which separate the larger particles greater than 10 to 20 micrometers by centrifugal force from the smaller particles, and fluidized bed separation, which removes smaller particles at the top (less than 50 micrometers) in the countercurrent overflow in a vertical column by gravimetric settling, and flotation, which is based on the different surface characteristics of contaminated particles. Addition of special chemicals and aeration in the latter case causes these contaminated particles to float. Screening is most applicable for particles larger than 1 mm. Magnetic extraction has not been successful for sediments. If the solids content is high, mechanical screening can be used. Gravity separation or sedimentation is applicable if the contaminated fraction has a higher specific gravity that the rest of the sediment fraction. According to the U.S. Army Engineer Detroit District (USACE Detroit District, 1994), costs are in the range of US$30 to US$70 per cubic meter for quantities in the range of 7,600 to 76,400 m^3 and for sediments with 75% sand and 25% contaminated silt or clay. The expense is only justified if the sediment contains more than 25% sand which is rare (National Research Council, 1997). Physical techniques only concentrate the contaminants in smaller volumes and are thus useful before thermal, chemical, or other processes.

The amount of dewatering depends on the type of sediments, dredging method used, and the technology for treatment. Mechanically dredged sediments typically contain more than 50% water (i.e., water content of more than 100%), whereas hydraulically dredged sediments contain more water than that. Centrifuges, filter presses, plate or diaphragm-plate filters, or gravity thickening can be used for dewatering purposes. Figure 7.2 illustrates the classification of mechanical dewatering for dredged materials being used in Japan.

In a demonstration project in collaboration with Environment Canada (1995), metal-contaminated sediments were removed from the Port of Sorel in the St.

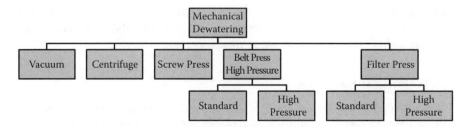

FIGURE 7.2   Classification of mechanical dewatering processes.

Lawrence River. The sediment was dewatered and treated with a rotary press and additives. Although this process removed 30% of the metals, which was sufficient for sediment disposal, it also added 30% to the cost of dredging and disposal. Various pilot and full-scale demonstration and commercial treatment processes have been developed and will be discussed in the following sections.

In Japan, similar techniques and processes are used to obtain aggregates from soils for concrete. The soils taken from mountainous areas are washed. The fine and light fractions are separated from coarse particles (concrete aggregates) in a centrifugal tank, and dewatered using the filter presser or belt presser. The water content is usually controlled at about 40%, depending on the energy cost and treatability of the materials. The technology can also be used for dredged materials.

7.3.2   SEDIMENT WASHING

Washing has been suggested for a variety of soils and target materials to be removed (Clarke et al., 1991; Davis et al., 1998; El-Shafey and Canepa, 2003; Kuhlman and Greenfield, 1999; Semer and Reddy, 1996). Sediment washing is a process that uses physical and/or chemical techniques to separate contaminants from soil and sediments (Interstate Technology and Regulatory Cooperation, 1997; http://www.itrcweb.org/). Figure 7.3 illustrates a typical soil washing process, where the separation consists of washing, rinsing, size separation, etc. Surfactant may be added in the washing water. Washing water and additives can be recycled or regenerated or treated prior to disposal. The dewatering of particles is needed. Mechanical dewatering, such as a filter press, conveyer filtration, centrifugal separation, etc., is available. The disposal of fine particles is different, depending on the type of contaminants and the contamination level. There are regulations for the disposal of contaminated sediments.

Washing processes generally use hot water. The viscosity of hydrocarbons is influenced by temperature, and the increase in temperature reduces the viscosity. Since the increase in temperature of water increases the kinetic energy of water molecules, the diffuse double layer of sediment particles becomes thinner. Therefore, surface attractive forces on the particles are reduced. The increasing temperature increases the solubilities of metal salts.

A typical washing method may be acid leaching, which refers to the remediation of sediment through extracting the metals with sulfuric acid. It is frequently not effective for cadmium. There are abiotic leaching and microbial leaching (Löser

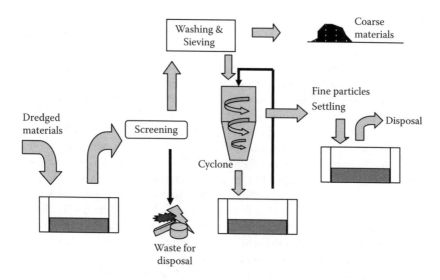

FIGURE 7.3 Size separation and washing of dredged materials.

et al., 2006, 2007). In abiotic leaching, the H_2SO_4 is supplied to the sediment. In this case, circulating water can be used. However, in microbial leaching, elemental sulfur is added to the sediment and is oxidized to sulfuric acid (Tsai et al., 2003a,b), or organic acids (such as citric acid produced by the fungus *Aspergillus niger*) for complexation of heavy metals may also be produced (Mulligan and Kamali, 2003). Both methods achieved removal efficiencies of greater than 90% for total extractable heavy metals. The pH ranges are controlled depending on the heavy metal species. For example, Al is markedly solubilized at pH < 4, and Fe at pH < 2.4.

Sediment washing involves the addition of a solution with the contaminated sediments to transfer the contaminants from the sediments to the wash solution. It is most appropriate for weaker bound metals in the form of hydroxides, oxides, and carbonates. Mercury, lead, cadmium, copper, nickel, zinc, and chromium can be removed and can be recovered by electrochemical processes if organic compounds are not significant. Metals can also be removed from precipitation or ion exchange. Precipitation is not applicable for metal sulfides. Pretreatment to remove uncontaminated coarser fractions can be used. Various additives can be employed such as bases, surfactants, acids, or chelating agents. Nitric, hydrochloric, and sulfuric acids can be used. However, if sulfuric acid is used, 50% of the amount is required compared to hydrochloric acid (Papadopoulos et al., 1997). The treated sediment can then be washed to remove any residual wash solution prior to disposal. Ideally the wash solution should be reused. Costs of sediment washing are usually in the order $40 to $250 per tonne (Hazardous Waste Consultant, 1996). Washing is usually most applicable for coarser particles. Therefore, fine-grain sediments can be difficult to decontaminate through washing solutions. Extraction tests should be conducted to determine optimal conditions (chemical type and dosage, contact time, agitation, temperature, and extraction steps to meet regulatory requirements).

Two companies, Biogenesis and Roy F. Weston, have combined mechanical and chemical processes for the removal of 90% organics and 70% inorganics (Amira et al., 1999). A full-scale facility was built to process 210,000 m^3/yr at a cost of $39 to $65 per m^3. Large facilities with capacities greater than 382,000 m^3/yr have been established.

The feasibility of using biodegradable biosurfactants to remove heavy metals from an oil-contaminated soil from a harbor area was recently demonstrated by batch washes with surfactin, a rhamnolipid, and a sophorolipid (Mulligan et al., 1999). The soil contained 890 mg/kg of zinc, 420 mg/kg of copper, with a 12.6% oil and grease content. A series of five batch washes removed 70% of the copper with 0.1% surfactin/1% NaOH, while 4% sophorolipid/0.7% HCl was able to remove 100% of the zinc. The results clearly indicated the feasibility of removing metals with the anionic biosurfactants tested, even though the exchangeable metal fractions were very low. These biosurfactants were also able to remove metals from sediments (Hall, 1998). Since these agents are biodegradable, they can enhance hydrocarbon removal and can potentially be produced in the sediments. The surfactants can be added as liquid or foam solutions (Wang and Mulligan, 2004).

It has been reported that surfactant can effectively remove the metals adsorbed on sediment particles. When a rhamnolipid biosurfactant without additives was applied, the removal of heavy metals from sediments was up to 37% of Cu, 13% of Zn, and 27% of Ni (Dahrazma and Mulligan, 2007). A scanning electron microscope (SEM) (JEOL JSM-840A) was used to examine the sediment. The washing tests were performed in a continuous flow configuration with a flow rate of 0.5 mL/min. SEM was performed for four samples. The samples were:

- Sediment before washing
- Sediment after washing with 2% rhamnolipid
- Sediment after washing with 1% NaOH
- Sediment after washing with 2% rhamnolipid and 1% NaOH

Pictures were taken with the SEM and shown in Figure 7.4. These pictures provide a general view of the fabric structure of the sediment. Comparing the pictures of the sediment samples before and after washing with 2% rhamnolipid (Figure 7.4(a) and 7.4(b)) shows that the textural structure of the sediment remains the same during and after the washing processes. In other words, the use of the rhamnolipid does not affect the natural size distribution of the sediment. This is an advantage of rhamnolipid as a washing agent. It also can be added that removal by rhamnolipid is an environmentally safe sediment treatment technique in both ex situ and in situ soil remediation. The sediment after washing can be returned to the environment with minimal damage to its natural texture in ex situ remediation, while the soil can remain in its natural place and texture for in situ sediment treatment. Adding 1% NaOH, according to Figure 7.4(c) and 7.4(d), decreased the particle size of the sediment. Adding NaOH dissolves the organic matter of the particles, which is larger than the carbonate and oxide fractions in the sediment and which complexes heavy metals.

The continuous-flow removal tests were performed for the sediment with different hydraulic conductivities at a flow rate of 0.5 mL/min with 2% rhamnolipid for 3 days. The relationship between hydraulic conductivity and removal for the metals is shown

(a)

(b)

(c)

(d)

FIGURE 7.4 Scanning electron micrographs of sediment at 1000× magnification. Washing tests were performed in a continuous flow configuration at the flow rate of 0.5 mL/min. (a) Sediments before washing, (b) sediments after washing with 2% rhamnolipid, (c) sediments after washing with 1% NaOH, and (d) sediments after washing with 2% surfactant and 1% NaOH (Dahrazma, 2005).

in Figure 7.5(a) and 7.5(b) for copper and zinc, respectively. By means of Equation 3.2, for a known fluid and specific soil sample, S_w can be determined if the k is known for any specific configuration. In both cases, the reduction of wetted surface area decreased removal of the metal from the sediment. This shows that the wetted surface area is among the parameters that control the mechanism of metal removal and is an important issue in continuous flow configuration where a heap leaching process might be performed. In addition, the removal of copper is more sensitive to wetted surface area, because the majority of copper in this sediment exists in the organic fraction. Organic materials have the largest surface area among all the fractions in the sediment. A decrease in the wetted surface area affects this fraction more than the others and consequently reduces the copper removal from the sediment.

Another plant-based biosurfactant (saponin) has also been evaluated. Rhamnolipid at a 2% concentration (pH 6.5), and saponin of 30 g/L (pH 5) were used to treat the sediments (Mulligan et al., 2007). Water alone (pH 5, 5.6, and 6.5) was used as the control in the experiments. In Figure 7.6, it can be see that after five washings of the

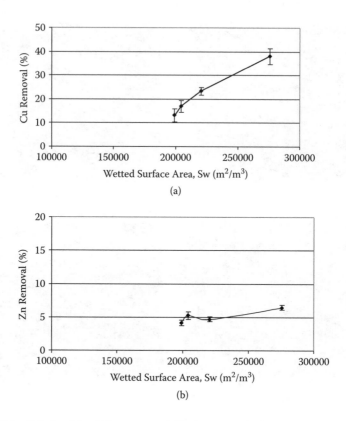

FIGURE 7.5 Relationship of (a) copper and (b) zinc removal with wetted surface area; 2% rhamnolipid with 0.5 mL/min in a continuous flow configuration was used for washing the sediments.

soil, rhamnolipids removed 46% of copper, 19% of zinc, and 10% of nickel, while the control removed 5% nickel and no percentage of zinc and copper. The rhamnolipids seems to have more affinity for copper than for zinc and nickel, as shown by the high removal rate. This phenomenon was also observed by Dahrazma (2005) where rhamnolipids removed more copper than zinc and nickel from sediment samples in a batch-washing test. Multiple washings appeared to improve the removal of the metals significantly, especially the removal of copper. The same trend was seen for the sediment sample, where up to 48% of the copper and 13% of the zinc were removed (Figure 7.6).

Saponin at a concentration of 30 g/L (pH 5) was used in a series of washings as shown in Figure 7.7. The saponin at pH 5 was able to remove 79% of zinc from the soil; however, the case was different for copper where the removal was 28% after five washings. The control (water) removed minimal amounts of all three metals (3.2% of zinc, <0.1% of copper at pH 5). It is also evident from Figure 7.7 that more than one washing of the soil improved the removal efficiency of the metals. Saponin seems to have a stronger affinity for zinc than for copper from the soil. However, in terms of percentage for the sediment, it is different, where 43%

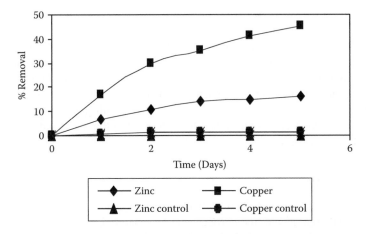

FIGURE 7.6   Removal of metals with 2% rhamnolipid at pH 6.5 from sediments.

removal was achieved for copper compared to 33% for zinc (Figure 7.7). There is a significantly higher initial amount of zinc (4,441 mg/kg) in the sediment compared to 894 mg/kg in the soil, which could account for the difference for zinc. In both cases, the trend for removal is the same, where more washes could potentially remove higher amounts of zinc, but copper seems to reach a static level after two or three washes.

Sequential extraction tests were performed on the sediments before and after washing with the controls and the washing agents. It was shown that copper could be removed mostly from the organic-bound fraction from the sediment and zinc and nickel from the oxide and carbonate-bound fraction by 2% rhamnolipid (pH 6.5). Saponin was effective for removal of heavy metals from all fractions with the exception of residual. Residual fractions, the most difficult to remove, were not

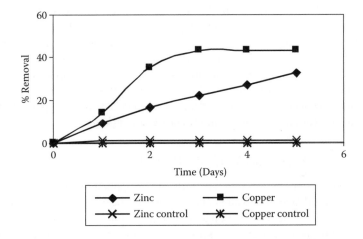

FIGURE 7.7   Metal removal from sediments with saponin (30 g/L) at pH 5.

affected during the surfactant washing studies and thus could be considered stable and unlikely to leach metals and unnecessary to remediate.

In Figure 7.8, the results are shown for the copper removal from the sediments. Apart from the residual, the organic fraction was the only significant fraction. The rhamnolipid was able to completely remove the copper from this fraction and, like the saponin for the soil, provided a complete treatment. The saponin removed only some of the organic fraction. The controls did not have any significant effect. Therefore the sequential extraction tests were useful for determining the metal binding fractions removed during washing.

7.3.3 Flotation

Flotation is a separation method of hetero-phase systems as dredged sediments. Contaminated organic and fine particles can be separated from the sediments using gaseous bubbles (Fujikawa, et al., 2007a,b). The flotation technique can also be used as in situ separation of sediments (Figure 7.9).

It is expected that various metal ions are adsorbed onto fine inorganic particles and retained by organic matter. Some metals would be present as sulfides in the dredged anaerobic sediments. The removal efficiencies for most heavy metals were up to 80% by a flotation process for sediments (Vanthuyne et al., 2003). Since the removed materials are not heavy metals, but fine particles and organic matter, other contaminants adsorbed on the removed particles must also be removed. For ex situ remediation of dredged materials, any variation, such as oxidation (see next section) or washing with or without (bio) surfactant, can be also employed at the same time, if necessary (Mulligan et al., 2001).

7.3.4 Ultrasonic Cleaning

Ultrasonic cleaning employs shear forces to remove materials attached to a surface (Meegoda, 2008). The shear force named cavitation is due to high energy acoustic cavitation. Bubbles are formed, grow, and then collapse. During this collapse, localized hot spots of 5000°C and 50,000 kPa occur for a few microseconds. The shock waves from cavitation induce interparticle collisions. The flow sheet of the process is shown in Figure 7.10. Bench-scale ultrasonic tests showed greater than 98% removal for polycyclic aromatic hydrocarbons (PAHs) and 95% for removal of chromium. For commercial treatment, the complete process included removal of coarse sediment by ultrasound followed by treatment of the fine and bulk solution by ultrasound and acoustic and flow fields for removal of fine sediments. The third step would include a membrane process to separate contaminants from the water.

7.4   CHEMICAL/THERMAL REMEDIATION

Most ex situ remediation techniques of soils can be applied for sediments. However, the existence of salt and high water content (200 to 500%) of sediments may become problematic to apply the remediation techniques. Chemical remediation includes the use of amendments, oxidation (also known as Fenton's reaction), and electrochemical remediation.

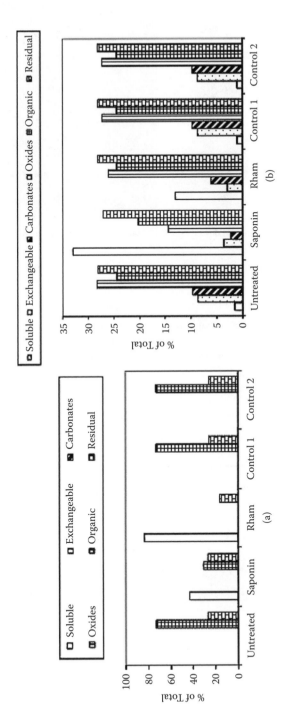

FIGURE 7.8 Sequential extraction of (a) copper and (b) zinc. Sap 1 = 30g/L saponin at pH 3; Sap 2 = 30g/L saponin at pH 5; Rham = 2% rhamnolipids at pH 6.5; Control 1 is water at pH 5, and control is water at pH 6.5.

(a) Illustration of experimental device.   (b) Photo showing particle flotation by bubbles.

FIGURE 7.9   Flotation separation of sediments.

7.4.1 OXIDATION

The oxidant, known as Fenton's reagent, destroys a variety of wastes and generates no harmful byproducts. Fenton's reagent was invented by Fenton in 1894. Today there are several methods known as "modified" Fenton's reactions, where different additives increase the oxidizing efficiency by increasing the pH tolerance, increasing the reaction time, and producing more and more stable radicals.

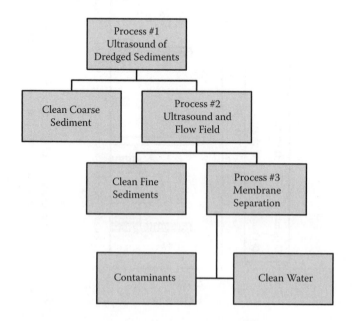

FIGURE 7.10   Ultrasonic process developed by Meegoda (2008).

$H_2O_2 + OH^- \ HO_2^- + H_2O$ (perhydroxyl radical)
$HO_2^- \ H^+ + O_2^{2-}$ (superoxide radical anion)
$HO_2^{\bullet} + O_2^{2-} \ HO_2^- + O_2$ (hydroperoxide anion)

The coexisting oxidation–reduction reactions associated with a modified Fenton's process promote enhanced desorption and degradation of recalcitrant compounds (Fenton, 1893, 1894, 1895; Fenton and Jackson, 1899). These include compounds such as carbon tetrachloride and chloroform, which were previously considered untreatable by Fenton's chemistry. There is a complete mineralization of organic matter. The breakdown is fast—within days, typically minutes–hours, depending on the concentration of H_2O_2. The process has some effects on the residual free phase.

Modified Fenton's reagent, hydrogen peroxide, and potassium permanganate were applied to sediments contaminated with PAHs (Ferrarese et al., 2008). Ferrarese et al. (2008) concluded that the optimal oxidant dosages determined were quite high, because sorbed PAH mineralization requires very vigorous oxidation conditions, especially for soils and sediments with high organic matter content. Their results indicated that the optimal oxidant dose must be carefully determined under site-specific conditions. Kellar et al. (2009) have used an in situ and ex situ application of a sodium-based Fenton reagent in the United States. The method has been proposed for the remediation of sediments near a chlorinated solvent site in Pennsylvania.

Wet air oxidation requires high temperatures and pressures, but is capable of destroying polycyclic chlorinated biphenyls (PCBs) and PAHs. Large quantities of water are not detrimental to the process. Costs are high at large scale, however.

Oxidation/reduction of heavy metals is another method for remediating ex situ sediments. A detoxification technology called TR-DETOX involves the percolation of inorganic and organic reagents to reduce heavy metals to their lowest valence state and form stable organometallic complexes. One of the main chemicals is sodium polythiocarbonate that forms a precipitate that becomes less soluble over time. The treated residue is no longer leachable. Lime, silicates, and Portland cement are not added, and costs are usually about one-quarter of stabilization/solidification processes. A unique characteristic is electronic addition of reagent. Pilot tests are required to determine the most appropriate formulation (Mulligan et al., 2001).

7.4.2 ELECTROKINETIC REMEDIATION

Electrochemical remediation uses a low DC current or a low potential gradient to electrodes that are inserted into the contaminated sediment (Virkutyte, 2002). When DC electric fields are applied to the contaminated sediment, ions migrate toward the corresponding electrodes (Figure 7.11). Cations are attracted to the cathode, and anions move to the anode. An electric gradient initiates movement by electromigration (charged chemicals movement), electroosmosis (movement of fluid), electrophoresis (charged particle movement), and electrolysis (chemical reactions due to electric field) (Rodsand and Acar, 1995). For example, under an induced electric potential, the anionic Cr(VI) migrates towards the anode, while the cationic Cr(III), Ni(II), and Cd(II) migrate towards the cathode. Once the remediation process is over, the contaminants that are accumulated at the electrodes are eventually extracted by

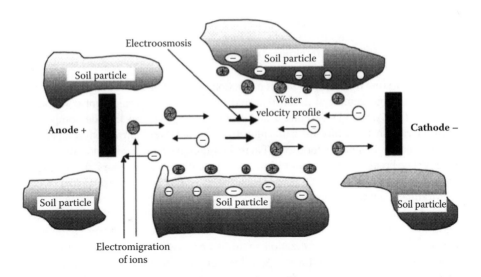

FIGURE 7.11   Electrokinetic remediation of sediments.

methods such as electroplating, precipitation/coprecipitation, pumping water near the electrodes, or complexing with ion-exchange resins (Reddy et al., 2001). The electric conductivity is the highest in the fine particles of the sediment on which also most metals are adsorbed, and the electric field is the strongest where the metals are mainly found. This method is well suited for fine-grained dredged sediment. In electrochemical remediation, there are four mechanisms, namely electromigration, electroosmosis, electrophoresis, and diffusion, affecting the migration of metals in an imposed electric field.

Control of the pH and electrolyte conditions within the electrode casings is essential in the optimization of the process efficiency. The process can be used to recover ions from soils, muds, dredging, and other materials (Acar et al., 1993). Dredged material is treated in lagoons between 2 and 7,400 cubic meters with batch times of 8 hours to 5 days, depending on current loading and electrode spacing. Metals as soluble ions and bound to soils as oxides, hydroxides, and carbonates are removed by this method. Other nonionic components can also be transported due to the flow. Unlike soil washing, this process is effective with clay soils.

Demonstrations of this technology have been performed, but are limited in North America (Mulligan et al., 2001). In Europe, this technology is currently used for copper, zinc, lead, arsenic, cadmium, chromium, and nickel. In the United Kingdom, it was evaluated for treatment of high concentrations of mercury in canal sediments. Other ions such as cyanide and nitrate and radionuclides such as uranium and strontium can also be treated by electrokinetics. Large metal objects, moisture content, temperature and other contaminants can interfere with the process. Recently, new developments at the pilot stage have been made in using electrokinetics for high-level metal-containing sediments. Metal recovery will improve the process economics.

Electromigration is the transport of ions and ion complexes to the electrode of opposite charge, while electroosmosis is the movement of soil moisture or groundwater from the anode to the cathode of an electrolytic cell. On the other hand, electrophoresis is the transport of charged particles or colloids under the influence of an electric field; contaminants bound to mobile particulate matter can be transported in this manner (Virkutyte et al., 2002).

Since sediment particles have a buffer capacity, they can release adsorbed substances from the surfaces when the value of pH decreases. Therefore, acidification may be a very effective method to solubilize the metal hydroxides and carbonates, other species adsorbed onto sediment particles, as well as to protonate organic functional groups (Yong and Mulligan, 2002). Generally, in the electrochemical remediation process, the development of an acidic front is often coupled with a successful remediation (Nystroem et al., 2006). However, because of the higher buffering capacity of sediments, acidification of dredged materials may not be an acceptable method. Surfactants can increase the solubility and mobility of heavy metals during electrochemical remediation, depending on its function on decreasing the ζ potential of sediment and then reducing the Van der Waals interactions (Nystroem et al., 2006). Therefore, using surfactants improves metal removal (Abidin and Yeliz, 2005).

7.4.3 SOLIDIFICATION/STABILIZATION

The purpose of solidification/stabilization processes is to reduce the mobility of the heavy metal contaminants by addition of an agent that solidifies and then immobilizes the metals. Cements, binders, and pozzolans are added (USEPA, 1994).

Solidification/stabilization is effective for metal contamination because there are few destructive techniques available for metals. Some metals such as arsenic, chromium (VI), and mercury are suitable for this type of treatment. Liquid monomers that polymerize and cement are injected to encapsulate the soils. Leaching of the contaminants must, however, be carefully monitored, as is the case for vitrification, the formation of a glassy solid. Cement- or silicate-based (5% to 10% by weight additives) processes are useful for sediments and are economical. Other materials containing iron (red mud, sludge from water treatment plants, bog iron ore, unused steel shot, and steel shot waste) have been evaluated (Mulligan and Kamali, 2003) for immobilizing cadmium and arsenic contaminants in sediments. All were effective in reducing the bioavailability of the metals to plants, but the safest was sludge from a drinking water plant with low levels of As. However, if there are different types of metals present, the treatment may not be as effective. Water contents greater than 20% or chlorinated hydrocarbons contents greater than 5% increase the amount of agents required. Variability in the water content, grain size, and the presence of debris can make handling of the materials difficult and decrease the efficiency of the solidification process. In addition, since immobilization leads to an increase in volume, larger areas of land are required for disposal. Thus, smaller volumes for treatment are more appropriate. Costs range from $30 to $250 per tonne (Hazardous Waste Consultant, 1996).

Full-scale projects have been performed in the United States, Canada, Japan, and Belgium. Halogenated semivolatiles, nonhalogenated semivolatiles and nonvolatiles, volatile and nonvolatile metals, low-level radioactive materials, corrosives, and cyanides have been treated effectively. In the Netherlands, a rotating drum was used in a full-scale experiment (Rienks, 1998). 680 tonnes of dewatered sediment were treated at 600°C for 38.5 hours for mineral oil, PAHs, and mercury. Mercury levels decreased by 80% from 1.5 to 0.3 mg/kg, while mineral oil and PAHs decreased by greater than 99.8%. Leaching of arsenic, molybdenum, and fluoride increased after thermal treatment, which can have implications in the reuse of the treated sediments as road or construction materials.

7.4.4 VITRIFICATION

Another immobilization technique is vitrification, which involves the insertion of electrodes into the soil, which must be able to carry a current and then to solidify as it cools. Toxic gases can also be produced during vitrification. Some vitrification processes have been tested on sediments. Costs can be high, because fuel values are low and moisture contents are high (above 20%).

A technology was developed for the remediation of organic contaminants and immobilization of metals in a glassy matrix and evaluated on the dredged sediments from New York/New Jersey Harbor (Institute of Gas Technology, 1996). A plasma torch is used to heat the sediments. Feeding of the wet sediments into the plasma reactor and adjustment of residence times can be difficult, however. Cadmium, mercury, and lead levels were reduced efficiently (97%, 95%, and 82%). Glass tiles and fiberglass materials were produced and could be used as valuable end products.

Temperatures higher than about 1200°C possibly degrade organic compounds and volatilize heavy metals. Because the solids like minerals will melt at this range of temperature, the technique which utilizes this temperature range is called vitrification technique or GeoMelt process. The materials can be burned, electrically melted, etc.

The GeoMelt processes are designed to be a mobile thermal treatment process that involves the electric melting of contaminated soils, sludges, or other earthen materials and debris, either in situ or ex situ, for the purpose of permanently destroying, removing, and/or immobilizing hazardous and radioactive contaminants. The ex situ technology for vitrification is illustrated in Figure 7.12.

Dredged materials are first dried and transported into a forge. The materials are melted at a temperature higher than 1200°C. The produced gas is cooled down and treated with activated carbon. After contaminants are removed by the activated carbon, the gas is released into the air. Since hazardous materials, such as organic compounds and heavy metals, in the materials are vaporized, the solids after vitrification are usually clean. This technique is recommended as one of treatment techniques of sediments contaminated with dioxin, in the Japanese technical guideline of sediments contaminated with dioxin. The process flow is shown in Figure 7.13.

Rotary kiln incineration has been used to produce cement. The technique has been applied to treat waste, contaminated soils, and sediments. The heat is supplied with a burner in a kiln, as shown in Figure 7.14, and the materials can be carbonized. Because the contaminants can be released by vaporization, flue gas treatment

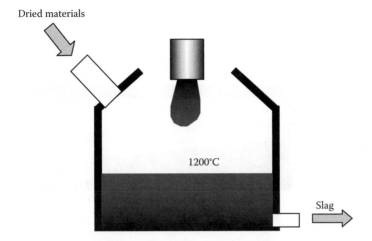

FIGURE 7.12   Vitrification of sediment.

is required. In Japan, this rotary kiln technique is also recommended as one of the treatment techniques for sediments contaminated with dioxin. PCBs can also be treated. The flow sheet of the process is shown in Figure 7.15. Costs, however, can be high.

There are soils contaminated with hydrocarbons in many industrial sites and oil refineries. Several technologies can be used for the remediation of these sites. Thermal treatments are the most popular and versatile techniques, because they can

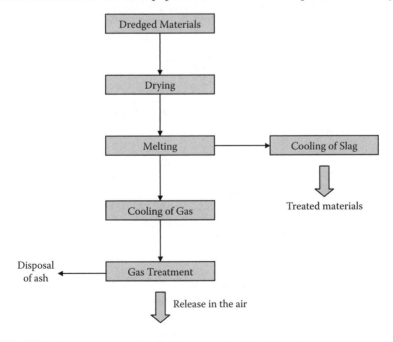

FIGURE 7.13   Japanese process for dioxin-contaminated sediments.

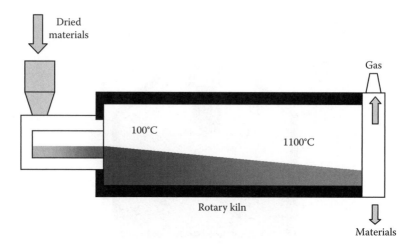

FIGURE 7.14 Rotary kiln treatment of sediments.

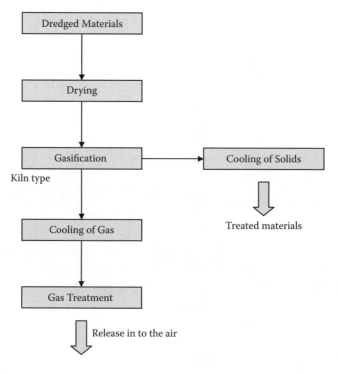

FIGURE 7.15 Flow sheet for rotary kiln technique for sediments contaminated with dioxin.

be effectively applied to a wide range of organic contaminants (Merino and Bucalá, 2007). Mechati et al. (2004) have studied the thermal desorption process using an industrial pilot-scale unit.

Merino and Bucalá (2007) performed experimental remediation on soils with hexadecane and investigated desorption and destruction temperatures. The results obtained at different temperatures (150 to 800°C) showed that at about 300°C the hexadecane can be removed almost completely from the soil matrix (99.9% destruction removal efficiency, DRE), and that temperatures above this value do not improve the removal efficiency noticeably.

In Japan, treatment of dredged materials is required for highly contaminated sediments with dioxin levels higher than 3000 pg-TEQ/g. The recommended techniques are as follows:

1. Melting (Geo-melting technique) (>1200°C)
2. Incineration (rotary kiln incinerator) (1100°C)
3. Low-temperature thermal degradation (400 to 600°C)
4. Chemical decomposition (350°C)

These techniques have various advantages and disadvantages.

7.4.5 THERMAL EXTRACTION

Mercury, arsenic, and cadmium and its compounds can be evaporated at 800°C with the appropriate air pollution control system. Some of the metals remain in the solid residues and will have to be properly disposed of. Thermal extraction is applicable mainly for mercury since this metal is highly volatile. Costs are in the order of $35 to $1000 per tonne (Environment Canada, 1995). However, there are numerous problems related to the treatment of sediments. Often the equipment is not appropriate for the feed size and moisture content of sediments. There are numerous suppliers, however, that claim that their equipment will work for the treatment of sediments.

There are several commercially available thermal chemical treatment processes. The temperatures used differ according to the process. Cement Lock, developed by the Institute of Gas Technology (IGT), has been used for dredged sediment in the New York/New Jersey Harbor (Stern et al., 1997). The sediment contained metal contamination (33 mg/kg As, 37 mg/kg Cd, 377 mg/kg Cr, 617 mg/kg Pb, 1.3 mg/kg Hg, 3.2 mg/kg Se, and 1.8 mg/kg Ag) and was fed with lime into the rotary kiln reactor smelter at 1200 to 1600°C. The mixture was then melted, and quenched, forming micrometer fibers. The mixture was then mixed with cement to produce a suitable type I Portland cement construction material. The sediment passed the toxicity characteristic leaching procedure (TCLP) for all metals. Volatilized heavy metals and acid gases and other combustion products were treated in the offgas by filtration to remove particulates and activate carbon to remove heavy metals gas. Based on pilot tests, costs were estimated at $20 to $30/m^3 (Rodsand and Acar, 1995). The pilot facility at Newark Bay, NJ, had a capacity of 23,000 m^3 per year. This type of process can be used for many types of dredged materials with no pretreatment. More recently a demonstration plant was completed in Bayonne, NJ, in July 2003 (Mensinger and Roberts, 2009). Two decontamination tests were performed between 2003 and 2007. Destruction and removal efficiencies (DRE) for the contaminants were on the order of 99.9%. The produced Ecomelt passed leachability tests and could be added for beneficial use in concrete, thus partially replacing Portland cement. Increased tipping fees could enhance the economics of the process and could lead to break-even scenarios.

Mercury Recovery Services (MRS) developed and commercialized a process that mixes a proprietary material and the mercury-contaminated material at temperatures of 150 to 650°C (Weyand et al., 1994). The process can be mobile or fixed, batch, continuous, or semicontinuous. Unit capacities ranged from 0.5 to 10 t/hour. The mercury can be as an oxide, chloride, and sulfide. No liquid or solid secondary products were generated. The treated material contained less than 1 ppm of mercury. The process consisted of two stages, feed drying and mercury desorption, which was then condensed as a 99%-pure metallic form from the vapor phase. Air emissions did not contain mercury. Costs were high, in the range of $650 to $1000/t.

The X-Trax™ process uses a relatively low-temperature process for removal of organics and mercury from soils, sludges, and sediments, which was developed by Chemical Waste Management Inc. The contaminated sediment is fed into a rotary dryer (400 to 650°C). Mercury was then desorbed. Oxide and sulfide forms were reduced to mercury. Nitrogen carries the vapors to the gas treatment systems. Approximately 10% to 30% of the mercury is removed by the dust scrubber. The scrubber liquid is then treated to separate water, organic, mercury, and sludge components. Nitrogen gas sent to a two-stage condenser enables mercury pure enough to be sent to an outside company. Droplets are removed from the gas by a mist eliminator. Approximately 5% to 10% of the gas is filtered to remove particulates in a particulate filter and carbon absorption system before discharge into the atmosphere, while the remaining amount is reheated and recycled to the rotary dryer. Soil and sediments with levels from 130 to 34,000 mg/kg of mercury have been reduced to levels of 1.3 to 228 mg/kg. Full-scale units can treat 10 t/h for sites with 20,000 to 100,000 t of contaminated soil (Palmer, 1996).

7.5　BIOLOGICAL REMEDIATION

For heavily contaminated sediments, various approaches can be used to enhance the rate of bioremediation. Substances must be biodegradable and not too toxic for treatment. Ex situ bioremediation has been more successful than in situ processes due to easier control of environmental parameters such as nutrient and oxygen contents because they can be added uniformly. Proprietary biological mixtures are also available. Ex situ biotreatment systems include the use of slurry bioreactors, biopiles, landfarming, and composting. The more sophisticated the process, the more expensive the treatment. Treatability studies are usually performed to determine the efficiency of the bioremediation for the type of contaminants and sediments at the site. The microbial population, nutrient levels, pH, moisture content, contaminant type, and concentration and sediment characteristics must be determined and followed. Bench-, pilot-, and demonstration-scale tests are needed to properly design the remediation technology.

Microorganisms have been effective in treating organic contamination in sediments such as PAHs. Zhao et al. (2004) have demonstrated that anaerobic degradation of RDX (hexahydro-1,3,5-trinitro-1,3,5-triazine) was possible in a Halifax sediment. Degradation rates of TNT > RDX > HMX (octahydro-1,3,5,7-tetranitro-1,3,5,7-tetrazocine) were found. *Shewanella* and *Halomonas* bacterial isolates were found (Zhao and Hawari, 2008). Khodadoust et al. (2009) showed that PCBs could be degraded anaerobically with the addition of iron periodically. The biodegradation of PCBs in lake and marine sediments was monitored for nine months, and the addition of ion in dosages between 0.01 and 0.1 g/g enhanced degradation.

7.5.1 SLURRY REACTORS

Slurry bioreactors use 5% to 20% solid content in a highly agitated treatment. Mass transfer, aeration, and environmental conditions can be optimized more easily than for in situ remediation. This type of treatment is particularly applicable for compounds of low biodegradability such as PCBs and PAHs. Slurry methods can be used because dewatering is not required (Figure 7.16). There are also other limitations as discussed for sediment washing. Bioremediation is a low-cost technology and therefore has the potential for wide use. However, metal remediation technologies are not as developed as organic treatments. Costs are in the range of $15 to $200 US per tonne (Environment Canada, 1995).

Surfactants can be added to enhance contaminant solubility, or the natural bacteria could be stimulated to produce natural biosurfactants. The latter approach was investigated for an oil and heavy metal-contaminated harbor soil (Jalali and Mulligan, 2008). It has shown potential and could be applied to contaminated sediments. Some preliminary results showed that, by the end of the 50-day experiment, nutrient amendments led to the enhancement of biosurfactant production up to three times their critical micelle concentration (CMC). Further experiments were performed to investigate the production of biosurfactants by limiting the inorganic source of nitrogen. Results showed an enhancement of biosurfactant production by 40%. The produced biosurfactants were also able to solubilize 10% of TPH and 6% of the metal content of the soil. These biosurfactants were produced by the indigenous soil microorganisms using organic contaminants as the sole carbon source. Furthermore, the produced biosurfactants showed potential to enhance biodegradation of petroleum hydrocarbons as well as to improve flushing of the remaining soil pollutants from the soil.

Baciocchi and Chiavola (2009) evaluated the use of a sequential batch reactor for treating sediment in a slurry phase. Degradation rates of PAHs of 90% to 95% could be achieved after optimization. Based on a 10% sediment concentration and laboratory tests, it was estimated that 4.8 kg/m^3/day could be treated.

7.5.2 LANDFARMING

Landfarming includes mixing the surface layer of soil with the contaminated sediment (Rittmann and McCarty, 2001) (Figure 7.17). Soil microorganisms are utilized for biodegradation of the contaminants. The resulting product is compost.

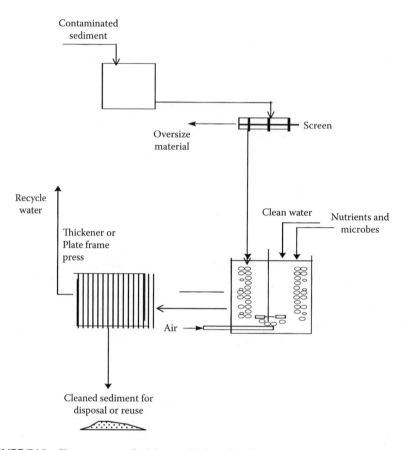

FIGURE 7.16   Slurry reactors for bioremediation of sediments.

Moisture must be monitored, and nutrients can be added to enhance biodegradation. Occasional turning of the soil increases the oxygen content and permeability of the sediment/surface soil mixture. The process is simple, but could lead to contaminant volatilization and leaching. Therefore, monitoring is required. Land requirements can be extensive. In the United States and Belgium, bioremediated dredged materials has been mixed with compost and/or municipal sewage sludge to produce soil for landscaping projects, and in Germany it has been used in orchards.

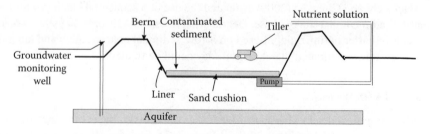

FIGURE 7.17   Landfarming process for bioremediation of sediments.

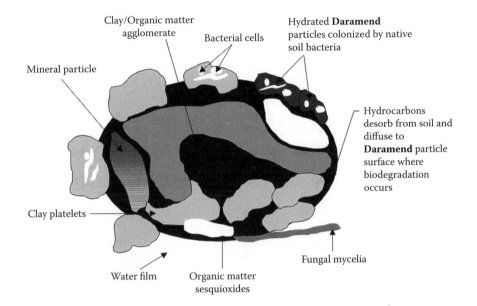

FIGURE 7.18   Schematic of the DARAMEND technology (from Mulligan, 2002).

An additive that has been used with landfarming is DARAMEND™. It is a solid-phase amendment (Figure 7.18) to promote anoxic conditions to enhance the bioremediation of pesticides such as toxaphene, DDT, DDD, and DDE. The reduction in the redox potential enhances the dechlorination of organochlorine compounds. With tilling equipment, the material can be mixed in to a depth of 2 ft. Hydrated lime is used to maintain the pH between 6.6 and 8.5. Redox potential and moisture were monitored and evaluated at a Superfund Site (Montgomery, AL) of a soil/sediment contaminated with pesticides (USEPA, 2004). Approximately 4500 tons were treated, and all contaminants reached specified levels. Santiago et al. (2003) evaluated DARAMEND for PAH-contaminated sediment. However, PAH concentrations were higher than expected (average of 900 ppm) and thus could not be reduced by bioremediation to CCME criteria (260 ppm) in bench-scale experiments. Thermal treatment was successful, however.

Metal removal can be accomplished in conjunction with organic removal. For example, Vega has developed a landfarming process that uses chelating organic acids with nutrients and soil conditioners to initiate biodegradation. The organic acids can chelate metals, as well as promote organic degradation. Temperature, moisture content, and pH need to be controlled as in any microbial process. It has mainly been applied for petroleum contamination. Retention times can be long (30 to 120 days).

7.5.3 Composting

Composting involves the biodegradation of organic materials to produce carbon dioxide and water. Typical temperatures are in the range of 55 to 65°C due to the heat from the biodegradation process. Animal or vegetable wastes such as sewage sludge are often used as organic amendments. Bulking agents such as wood chips are added to increase the porosity of the material. Moisture content and temperature must be monitored. Composting processes include windrows and biopiles and in-vessel composting.

Composting of a contaminated sediment was evaluated by Khan and Anjaneyulu (2006). A ratio of 10 kg of sediment with 0.5% fertilizer and 50% compost was used. The contaminants present in the sediment included phenols (16 to 24 mg/kg) and benzene (3.4 mg/kg). Fertilizer was added to increase the nutrient content, and compost was used as the inoculum of microorganisms. Wood chips were added as a support and aerating material in the pile for composting. The parameters, pH, total volatile solids, microbial count, temperature, and contaminant concentration, were monitored over the period of five weeks. Approximately 80% to 85% of the phenols were degraded, whereas benzene was almost completely biodegraded. Therefore, composting was shown to be technically feasible at lab scale.

Myers and Williford (2000) examined the bioremediation of contaminated sediments in a confined disposal facility (CDF). Composting (windrows and biopiles), landfarming, and land treatment were examined (Table 7.1). The contaminants included PAHs, PCBs, and PCDDs/Fs (dioxins). Land treatment is similar to land farming, except that the contaminated sediments are tiled and interact with the surrounding soil. Monitoring is essential due to potential leaching and volatilization of contaminants. Laboratory studies have shown the biodegradability of these compounds, but there is a lack of information on the treatment of dredged material. Composting and land treatment have potential to be cost effective, but require understanding of the biological processes and the technology. Pilot and demonstration studies are needed to do this. Subsequent composting tests were not successful in remediating PAHs. PCB degradation may be a little more promising. The factors and conditions were not well understood, and further research work is needed (Myers et al., 2003).

TABLE 7.1
Comparison of Bioremediation Technologies

Parameter	Windrow Composting	Landfarming	Biopile Composting
Applicability	Explosives, PAHs	Diesel fuel, fuel oil, PCBs, pesticides	Fuels, solvents
Site requirements	Excavation and special mixing equipment	Excavation and earthmoving equipment	Excavation and earthmoving equipment
Limitations	Bulking agents increase volume and may need to be removed	Permanent structures required	Static process without mixing
Cost	$248/m³	Less than $98/m³	$35 to 130/m³

Source: Adapted from Myers and Williford, 2000.

7.5.4   BIOLEACHING

Bioleaching involves *Thiobacillus sp.* bacteria which can reduce sulfur compounds under aerobic and acidic conditions (pH 4) at temperatures between 15 and 55°C, depending on the strain. Leaching can be performed by indirect means, acidification of sulfur compounds to produce sulfuric acid which then can desorb the metals on the soil by substitution of protons. Direct leaching solubilizes metal sulfides by oxidation to metal sulfates. In laboratory tests, *Thiobacilli* were able to remove 70% to 75% of heavy metals (with the exception of lead and arsenic) from contaminated sediments (Karavaiko et al., 1988).

Options are available for bioleaching, including heap leaching and bioslurry reactors. Sediments require lower pH values to extract the metals because they have already been exposed to oxidizing conditions. For both heap leaching and reactors, bacteria and sulfur compounds are added. In the reactor, mixing is used, and pH can be controlled more easily; leachate is recycled during heap leaching. Copper, zinc, uranium, and gold have been removed by *Thiobacillus sp.* in biohydrometallurgical processes (Karavaiko et al., 1988).

Percolation field tests were run by Seidel et al. (1998). They found that addition of sulfur as a substrate provided better leaching results than sulfuric acid. Approximately 62% of the metals were removed by percolation leaching after 120 days for the oxic sediments. Only 9% of the metals were removed from the anoxic sediments. They indicated that anoxic sediments are less suitable for treatment and must be ripened as a pretreatment.

7.5.5   BIOCONVERSION PROCESSES

Microorganisms are also known to oxidize and reduce metal contaminants. Mercury and cadmium can be oxidized, while arsenic and iron can be reduced by microorganisms. This processs (called mercrobes) has been developed and tested in Germany at concentrations greater than 100 ppm. Since the mobility is influenced by its oxidation state, these reactions can affect the contaminant mobility.

Chromium conversion is also affected by the presence of biosurfactants. A study was conducted by Massara et al. (2007) on the removal of Cr(III) to eliminate the hazard imposed by its presence of kaolinite. The effect of addition of negatively charged biosurfactants (rhamnolipids) on chromium-contaminated soil was studied. Results showed that the rhamnolipids have the capability of extracting a portion of the stable form of chromium, Cr(III), from the soil. The removal of hexavalent chromium was also enhanced using a solution of rhamnolipids. Results from the sequential extraction procedure showed that rhamnolipids remove Cr(III) mainly from the carbonate and oxide/hydroxide portions of the soil. The rhamnolipids also have the capability of reducing close to 100% of the extracted Cr(VI) to Cr(III) over a period of 24 days.

7.5.6   PHYTOREMEDIATION

Some plants have been shown to retain metals in their roots, stems, and leaves (Hazardous Waste Consultant, 1996). Vegetative caps consisting of grasses, trees, and shrubs can be established in shallow fresh water. The resulting vegetative mat can hold sediments in place. The construction of wetlands is growing for wastewater treatment, and thus the knowledge on wetland configurations is growing. However, vegetative caps have not yet been applied to the remediation of sediments (Mulligan et al., 1999). It is more likely that this technology will be used as an in situ method of reducing large volumes of sediment transport. However, phytoremediation could be implemented where dredged sediments have been placed in contained areas and a wetland is then constructed to remediate and contain the sediments. Lee and Price (2003) indicated that, although phytoextraction of Pb with chelates may be troublesome due to potential leaching into groundwater, immobilization and phytostabilization can be appropriate in CDFs. The site could potentially be restored for beneficial use as a wildlife habitat.

7.6   BENEFICIAL USE OF SEDIMENTS

There are two choices regarding the handling of dredged materials: beneficial use or disposal. For the use of dredged materials, it is problematic that dredged materials are too soft and contaminated. In addition, the volume of dredged materials is often very large. Since there is a lack of disposal sites, beneficial use of the contaminated dredged sediments is promoted. Between the alternatives, there may be various choices, as shown in Figure 7.19.

Beneficial use of dredged sediments has been investigated by many researchers (Sadat Associates Inc., 2001; Colin, 2003; Comoss et al., 2002; Dermatas et al., 2002; Douglas et al., 2005; Dubois, 2006; Maher, 2005; Maher et al., 2004, 2006; Siham et al., 2008; Yozzoa and Robert, 2004; Zentar et al., 2008).

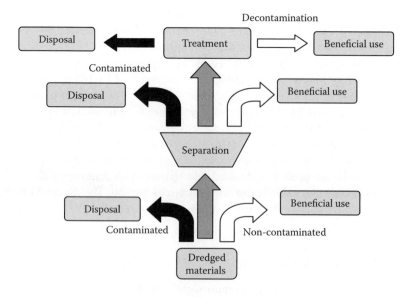

FIGURE 7.19   Reuse strategies for decontaminated sediments.

Many studies have been reported from the Great Lakes project (Great Lakes Commission, 2004).

It is necessary to improve the materials for use as construction materials. Zentar et al. (2008) investigated the mechanical behavior and environmental impacts of a test road built with marine dredged sediments. They improved the mechanical properties of fine dredged materials by adding dredged sand. Basically, the result can be dependent on the grain size distribution of the mixture (Fukue et al., 1986). The leaching test results showed no significant leachate production from the materials, because of the initially low concentration of toxic substances for the raw materials and high pH values used. In fact, sediments have been in water, and the amount of leachate during leaching tests is usually low (Fukue et al., 2001). It is suggested that leaching tests should be performed for various pH values, especially for lower pH values.

Contaminated dredged material is a problem worldwide. For this reason, the disposal of dredged materials into the ocean is prohibited (London Convention). Jones et al. (2001) compared various decontamination techniques used for the dredged sediments from the port of New York/New Jersey. In the report, a total of nine different technologies were introduced. Experiments were performed at bench scale (15 liters). The approaches included sediment washing, solvent extraction, thermal desorption, and thermal destruction. These technologies can be viewed as components of a treatment train for dredging, treatment, and beneficial use of contaminated dredged material. They also discussed beneficial use and commercialization of the products.

The sediment washing can lead to manufactured soils or material from residential landscaping (USEPA, 2005). The thermal treatment can produce manufactured grade cement comparable to Portland cement. The treatment train included dewatering, pelletization (a type of solidification by addition of shale fines and extrusion), and transportation to an aggregate facility. After pelletization, the pellets are treated in a rotary kiln, exploding the organic matter. The final product can be used for various geotechnical applications including concrete production and insulation of pipelines. The last process, vitrification, produced a glassy material that could be used for architectural tiles.

Beneficial use is not normally considered in treatment processes. However, it should be to enhance the sustainability of the process. Cost-effectiveness is a major consideration, and contaminant release must not occur. Besides those already discussed, alternative products can include:

- Construction fill
- Municipal landfill cover
- Restoration of mined areas
- Capping material
- Building materials
- Enhancing beach areas with clean sand
- Habitat restoration in dredged areas or wetlands

Artificial sand beaches and tidal flats are created for one of the following purposes:

1. Formation of clean beach for resort areas and parks
2. Farming for shellfish
3. Recovery of beach following reclamation
4. Rehabilitation of coastal marine environment

Sand beaches and tidal flats possess natural capabilities for cleaning seawater under repeating waves and tides. This capability arises from a combination of their ability to filter a large amount of suspended solids (mostly organic matter) and the dissolution of the suspended solids by microorganisms. Although the organic matter entrapped in the sand pores is food for microorganisms and benthic animals, there are no easy means to quantify the process and its benefits. Evaluation of the impacts arising from construction of the tidal flats and beaches cannot be readily performed. In part, this is due to the dynamic processes initiated by the actions of currents and waves. Stabilization of the new beaches and tidal flats will be a long-term process. The use of breakwaters on beaches brings with it problems of decrease in redox in the region due to the dead organisms and excrements. One of the three tidelands is Kansai Rinkai Park, with an area of 270,837 m^2, which was created artificially by the Tokyo Metropolitan Government in 1965, at a time when Tokyo Bay was losing its valuable natural environment. The area incorporates vast tidelands, which were once the breeding areas for birds and were also once abundant with fish and shellfish.

Geotextile tubes can be used to protect coastline from waves and tides—as has been utilized in some countries. The tubes, which can be installed along the coastline, are a few meters in diameter and a few kilometers in length and can be filled with dredged sediment. They can also serve as a breakwater for man-made islands and wetlands.

7.7 CONFINED DISPOSAL

In the case that sediment is contaminated, in situ/ex situ remediation can be performed. If remediation is not possible, the dredged materials can be disposed of in a proper manner. Containment of dredged material is carried out in confined disposal facilities in dikes near the shore, island, or on the land facilities. The facilities must be designed for dredging purposes and to contain the contaminants. Landfills have been used widely for disposal of dredged materials. The sediments must be previously dewatered such as in a contained disposal facility because landfill facilities cannot handle slurries. Large volumes cannot usually be accommodated because landfills do not have the capacity. Potential mechanisms for contaminant release are due to leachates, runoff, effluents, volatilization, uptake by plants, and ingestion by animals. Therefore, pretreatment by stabilization/solidification may be necessary. Oxygenation of sediments by the rain can lead to metal contamination of the groundwater. The cost is in the range of $20 to $65 per cubic meter (USEPA, 1993).

Containment facilities can be used for storage, dewatering, and pretreatment for other processes. These costs are usually less than those for landfill. Areas for contained aquatic disposal, the placement of material in a confined aquatic area called a confined disposal facility (CDF), can be strategically placed in depressions and confined by dikes. This technique can be used for disposal of contaminated sediments.

Clean material can be placed above and at the edges. The USACE and USEPA (2003) have reviewed the use of CDFs for dredging projects in the Great Lakes.

Confined aquatic disposal (CAD) is used for placement of dredged material in a natural or excavated depression. It has been used mainly for navigational purposes such as in Boston Harbor, not disposal of contaminated material. It may be appropriate if landfill disposal or in situ capping is not possible. Maintenance costs are low, and there can be an increased resistance to erosion. Depths can be a few to more than 10 m, and widths are in the range of 500 to 1500 m. As they are filled, capping is used.

Another approach is to place the material in woven or nonwoven permeable synthetic fabric bags, geotextile tubing, or containers (NRC, 1997). Costs at the demonstration in California were approximately $65 per m^3 (Clausner, 1996). The contaminants must not seep through the fabric into the water, and these uncertainties must be further investigated.

The U.S. Corps of Engineers have used geocontainers to store dredged sediments. The geocontainers are made of geosynthetic material and assembled by a seaming technique. Large quantities of dredged material are contained in the geocontainers after filling by hydraulic or mechanical filling equipment. This can be done in situ or in split-hull barges. If the latter is used, the sediments can be pumped as a slurry into the bags. This is followed by stitching of the bags and allowing them to fall through the split of the barge. The geocontainers dropped from barges into open water can form underwater berms, dikes, or other structures. They are designed to resist degradation under environmental conditions. The containers can be used (Rankilor, 1994) as breakwaters, near beaches or offshore, to stabilize sand dunes or wetlands or for dike construction.

In the Mississippi River near Baton Rouge, the Red Eye Crossing Soft Dikes Demonstration Project (Hall, 1998) used polypropylene bags filled with coarse river sand as soft dikes. Millions of dollars can be saved because less dredging is required. The soft dikes are placed lower than the nearby sandbar where the bags are filled. Both small geobags of three cubic meters and large geocontainers of 200 to 300 cubic meters are used. The project has gone well for over four years.

In Japan, dredged sediments are basically regarded as waste materials. Therefore, the dumping of dredged materials into the ocean is basically prohibited, and dredging cannot be achieved unless the site for disposal is ready. This is mainly because sediments are usually organic rich and more or less contaminated. There is one exception when dredging of sediments will be recommended. Because there is a guideline for dioxin-contaminated sediments, when the contamination of sediments with dioxin is found, dredging is one of the effective methods to solve the problem. The selective disposal flow sheet for dioxin-contaminated sediments is shown in Figure 7.20 (Japanese guideline for investigation, treatment and measure for dioxin contamination of sediments, Revised version, 2008). The fractions of contaminated dredged materials or highly concentrated dredged materials with toxic substances, which cannot be beneficially used, have to be disposed of without subsequent contamination.

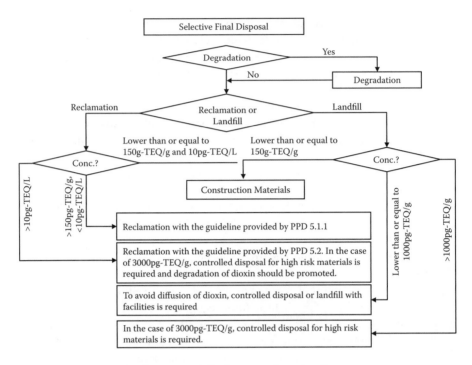

FIGURE 7.20   Selective flow chart of dioxin-contaminated sediment.

7.8   COMPARISON BETWEEN TREATMENT TECHNOLOGIES

A major problem in comparing treatment technologies is that very few studies use the same sediment. Recently, however, the EPA's Great Lakes National Programs Office performed a study on Trenton Channel sediments (Cieniawski, 1998). Five technologies including solid phase extraction, solidification (Growth Resources), soil washing (Biogenesis), thermal desorption (Cement Lock), and plasma vitrification (Westinghouse) were evaluated. A drum of 208 L of sediments was given to each company. The sediments contained PAHs, mercury, lead, PCBs, and oil and grease. Solid-phase extraction had no significant effect on total metals. High-temperature plasma vitrification was effective for greater than 90% of the contaminants including the metals. Conversion of the sediment into the form of glass will allow its use as an aggregate or glass tile or in glass fiber products. Cement Lock was very effective for all contaminants with the exception of metals (20% and 90% reduction). Heavy metals are locked in the cement matrix, while volatile metals, such as mercury and arsenic are volatilized. Volume reduction is a major advantage of the process. In addition the cement end product can be used in construction, eliminating disposal costs. Although soil washing was very effective for leachable metals, it was only partially effective for total metals. Wastes were reduced to reusable oil, treated water, and soil for backfill. Overall, most of the technologies were able to remove mercury but not lead to residential criteria. Only industrial and commercial criteria could be achieved for lead. Estimation of the costs for the technologies showed that the highest capital costs were \$10 to \$15 million for the vitrification process and \$20 million for Cement Lock, while those for soil washing were \$3.5 million, and for solidification \$0.7 million (Snell Environmental Group, 1997). Operating costs were the highest for soil washing (\$118/m^3), followed by vitrification (\$110/m^3). The lowest operating costs were for the thermal desorption (\$63/m^3) and solidification (\$59/m^3). Although these tests were performed at bench scale, they are useful in the comparison of technologies because they are on the same basis. Because Cement Lock and vitrification achieved the highest removal efficiencies and produced useful final products, they were recommended for further pilot tests. In 1999, it was decided to remove 23,000 m^3 of contaminated sediment from Black Lagoon and treat a fraction with Cement Lock (Zarull et al., 1999).

7.9 CASE STUDIES OF REMEDIATION

7.9.1 REMEDIATION OF SEDIMENTS CONTAMINATED WITH DIOXIN

In Japan, dioxin is designated as a special substance to be extremely toxic for humans, and the guideline provides that sediments are contaminated if the dioxin level is higher than 150 pg-TEQ/g. It was found that the sediments in Tagono-ura port in Fuji City, Japan were contaminated with dioxin (Table 7.2). Based on the regulations concerning sediments contaminated with dioxin, the sediments had to be remediated. The investigation prior to the project showed the contamination level and volume as follows.

The contaminated area was 349,000 m^2. The project was led by the committee established in the Shizuoka Prefecture. The committee examined three techniques (i.e., dredging, capping, and in situ solidification) and selected dredging and disposal. The committee selected dredging using the grab technique, because of low contamination in the surrounding area. During the dredging, a silt fence was used to avoid diffusion of contaminated particles.

The basic process from dredging to final disposal is shown in Figure 7.21. The treatment consisted of solidification, separation, and dewatering. Solidification prevents contamination from spreading and dioxin from leaching. Separation and dewatering was used to reduce the volume. Monitoring was made to inspect the contamination from the resuspended particles. Since measuring dioxin is expensive, the committee chose the following procedure.

TABLE 7.2

Contamination Levels and Volume of Sediments in Tagono-ura, Japan

Concentration (pg-TEQ/g)	Contaminated Volume (m^3)
150–1000	471 000
1000–3000	70 000
More than 3000	1000

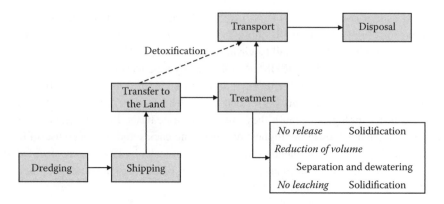

FIGURE 7.21   Basic process from dredging to disposal.

The relationships between the concentration of dioxin and turbidity in seawater were determined with different site locations. The relationships were linear, and correlation factors were very high. For example, the following relationship was determined.

Quality of seawater with dioxin C_D (pg-TEQ/L)

$$= 0.1533 \, T_b \quad (R^2 = 0.9986)$$

where C_D is the concentration of dioxin in seawater, and T_b is the turbidity (NTU). This shows that dioxin is adsorbed to the suspended solids. Using the relationship, C_D values were estimated from the measured turbidity. In fact, the relationships obtained were different for the different site locations. The relationships were expressed by:

$$C_D = kT_b$$

where k is a proportional constant.

The allowable level was decided as 1 pg-TEQ/L. Therefore, the following guideline was used for the control of seawater quality.

$$T_b < \frac{1}{k}$$

The T_b values may depend on rainfall or discharges from rivers. However, the higher the amount of rainfall or suspended solids (SS) discharged, the more conservative the T_b obtained will be. If there is a risk that the level of dioxin estimated closes to the allowable level, the dredging work is stopped. Until the cause of the high value is investigated and measured, the dredging work cannot be started.

7.9.2 DREDGING CASE STUDY

The environmental restoration of Cells 1 and 3 of sector 103 of the Port of Montreal (Vallée, 2008) was recently performed. The project was initiated in 1989 with the

TABLE 7.3

Characterization of the Sector 103 sediments from the Port of Montreal

Contaminant	Cell 1 Average Concentration (ppm)	Cell 3 Average Concentration (ppm)
C_{10}–C_{50} hydrocarbons	6,703	11,762
Phenanthrene	15	24.6
Copper	166	4770
Selenium	5.8	195

Source: Vallée, 2008.

characterization of the sediments. A working group was then formed by Environment Canada and included the Montreal Port Authority, Xstrate, Shell, and Imperial. Some preliminary studies were performed from 1994 to 2001, and a protocol was signed in 2001. The environmental impact study took place from 2001 to 2003 because the work must protect the public, the aquatic ecosystem, and the environment. Work finally started in 2007 and continued into 2008. The characterization is shown in Table 7.3. It was estimated that 40,000 m³ of sediment were contaminated.

The following steps were then carried out. Construction of infrastructure for sediment storage was followed by installation of silt curtains to confine suspended solid transport. Mechanical dredging was performed. Turbidity was monitored throughout the procedure. The level of suspended solids did not exceed 25 mg/L (50 m downstream) (MDDEP criteria) outside the silt curtain. Removed sediments were then transported for subsequent dewatering and storage of the sediments. Bathymetric tests were performed after dredging to minimize the residual contaminated sediments left in place. The sediments were then disposed of after drying. The initial water content was reduced to 38% in 7 to 10 days. The work in Cell 1 was started in 2008 and will be completed by 2012. In Cell 3, approximately 9,700 metric tonnes were dried at Shell and disposed of at Grand-Piles. Another 9,900 tonnes were removed in 2008. In Cell 3, the level of petroleum hydrocarbons was higher than estimated. Initially only 40% was estimated to be above the accepted level for C10–50 but more than 95% of the sediments was higher than relevant ecological screening criteria (RESC), demonstrating the difficulties involved in estimating the amount of contaminated sediments that needs to be removed.

According to the Group Restauration de 103 (www.grouperestauration103.com), more than 52,278 m³ of contaminated sediments were removed, which was more than the initial objective. More than 91% of the contaminants were removed as indicated in Table 7.4.

There was only one complaint regarding dust from the trucks involved in the sediment transport work. No infractions of the regulations occurred, including the air quality criteria of the Shell and Imperial oil companies. All water quality criteria were respected in the St. Lawrence River. No odor was detected by the residents in the area. Noise level limits of 75 dBA were respected and were maintained at 62 dBA. The managers of the project, through careful planning, avoided impacts to the

TABLE 7.4
Amount of Contaminants Removed from Cells 1 and 3 at Pier 103

Contaminants	Amount Removed (kg)	Removal (%)
Petroleum hydrocarbons	485,499	98.5
PAH	563	99.0
Copper	105,894	98.5
Selenium	2,320	98.5

Source: http://grouperestauration103.com

residents, environment, and aquatic ecosystem at the site. After dredging, more than 9,669 metric tons of sediments were sent to the authorized center for the treatment and confinement of contaminated sediments and soils at Grandes-Piles. For the sediments contaminated with hydrocarbons only, the soils are being treated by biotreatment at the Imperial site near the dredging area until 2012.

7.9.3   CASE STUDY OF A WASHING PROCESS

In Quebec, Canada, the Institut National de la Recherche Scientifique (INRS) developed the ORGANOMETOX process for removal of inorganic and organic contaminants (Dragage Verreault, 2008; Mercier et al., 2008). The process was developed for soils and sediments that were very fine because the contaminants were mainly concentrated in that fraction (Figure 7.22). The project was initiated in 1995, and the process is the subject of U.S., Canadian, and European patents (7 countries). In the pilot plant more that 40 tonnes of soils/sediments have been treated at a rate of 8 tonnes/hour. The process is indicated in Figure 7.23. The first step includes metallurgical processes such as screening, for obtaining the fine fraction with the highest level of contaminants. Surfactants are then added for organic contaminant removal followed by various gravimetric separation processes. The surfactant is cocamidopropyl hydroxysultaine (CAS) flotation columns and solid/liquid separation by centrifuge. The pilot tests can then be used to determine the technical economic feasibility of a commercial-scale plant of 50 tonnes/hour. Commercialization is difficult because the results must be guaranteed by the contractor. The cost is usually relatively costly (in the range of $100 to $300 per tonne). Few applications have been accomplished. For sediments from Montreal with initial zinc and copper contents of 2682 and 117 mg/kg, respectively, removal of each was 88% and 70%.

7.9.4   BIOTREATMENT CASE STUDY

A pilot field test was performed in a confined treatment facility (CTF) for PCB (Arochlor 1248 and 1254) contaminated sediments from the Sheboygan River in Wisconsin (Bishop, 1996). A 1,300-m² CTF was constructed of steel sheet piling. It was divided into four cells to include two treatment and two control cells. Water was discharged via a permeable wall. The cells included leachate control, and nutrients,

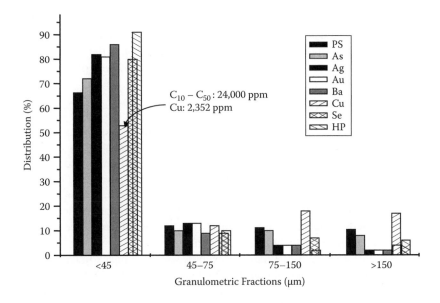

FIGURE 7.22   Characterization of sediments according to particle size (Mercier et al., 2008).

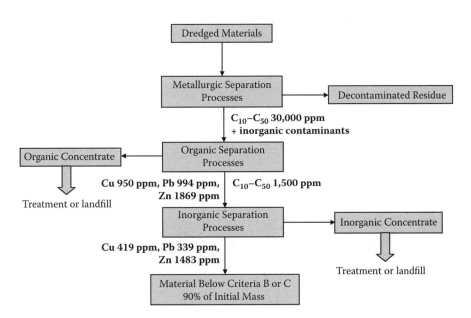

FIGURE 7.23   ORGANOMETOX process developed by INRS (Mercier et al., 2008).

oxygen, and other amendments were added. Both aerobic and anaerobic treatments of the cells were evaluated. Sufficient quantities of oxygen could not be supplied to the cells. Anaerobic conditions did not lead to dehalogenation of the PCBs. Sediment samples were extremely heterogeneous, making statistical analysis difficult.

7.10   CONCLUDING REMARKS

Ex situ remediation of contaminated sediments basically requires dredging for removal of the sediments. Dredged materials are either beneficially used or are disposed of, depending on the degree of contamination and grain size. To use the contaminated dredged materials, they often require remediation. The remediation technique for dredged materials may be the processes of washing, sorption, decomposition, or/and degradation. The techniques are not dissimilar to those used for soils. When dredged materials are obtained from the sea, the existence of salt may prevent use of some of the techniques established for soils.

The cost performance of the remediation processes is complex in terms of availability of disposal site, price of construction materials, remediation operation cost, and the international and domestic constraints in the environmental situation. The cost and price are variable with time and site characteristics. A summary is shown in Table 7.5. Ex situ bioremediation allows better control of environmental parameters such as pH, oxygen, and mixing than in situ bioremediation. However, there are still many challenges related to bioavailability of the contaminants. This issue can be remedied through the use of biological surfactants.

TABLE 7.5
Comparison of ex Situ Remedial Technologies Sediments

Technology	Description/Scale	Applicability	Effect on Contaminants	Waste Streams Generated	Siting Requirements	Beneficial Use	Costs ($US/ metric ton)
Stabilization/ solidification	Creation of an inert waste/full scale	Injection of solidifying chemicals	Immobilization within matrix	Debris from screening	Large area for mixing equipment, power	Compacted fill, capping material	60 to 290
Vitrification	Application of electrical energy to vitrify contaminants	Low volatility metals and production of glass materials/ pilot scale	Thermal oxidation and metal immobilization	Debris from screening and wastewater	Plasma arc facility/ dewatering, fuel, power	Glass aggregate for various uses	90 to 870
Physical separation	Includes froth flotation, gravity separation, screening, etc.	For high metal concentrations/ full scale	Removal of concentrates in small size fractions	Various solid and liquid wastes and debris	Power, separation equipment	Capping materials	Usually included with washing
Washing	Addition of surfactants and other additives to solubilize	For sand and gravel-sized sediments/bench and pilot scale	Metal and organic removal, possible organic oxidation	Debris from screening, wastewater, and sludges	Water, power, mixing equipment	Manufactured soil with addition of bulking agents	60 to 245

—continued

TABLE 7.5 (CONTINUED)
Comparison of ex Situ Remedial Technologies Sediments

Technology	Description/Scale	Applicability	Effect on Contaminants	Waste Streams Generated	Siting Requirements	Beneficial Use	Costs ($US/ metric ton)
Thermal	Elevated temperature extraction and processing for metal removal	Highly contaminated volatile metal-contaminated sediments/full scale	Volatilization of volatile metals, destruction of organics	Emissions of volatile organics and metals	Power, fuel/ chemical additives	Manufactured soil by addition of agents, fill, capping material	250 to 9,000
Biological leaching	Use of microbes for metal heap leaching or in slurry reactors	Applicable for sand and gravel-sized sediments/bench scale	Metal removal	Debris from screening, wastewater, and sludges	Power, water, slurry reactors	Manufactured soil with addition of bulking agents	15 to 200
Phytoremediation	Use of plants for metal extraction	Applicable for low metal contaminant levels but not yet demonstrated	Metal and some organics removal	Plant debris with contaminants	Little preparation needed	Manufactured soil	Little info
Electrokinetics	Use of electrodes to mobilize contaminants	Metal and organic contaminants/ bench scale	Metal and organic chemical removal	Limited residues, electrodes may require disposal	Power requirements	Manufactured soil	50 to 150

REFERENCES

Abidin, K. and Yeliz, Y. 2005. Zeta potential of soils with surfactants and its relevance to electrokinetic remediation. *J. Hazard. Mat.* 120: 119–126.

Acar, Y.B., Alshawabkeh, A.N., and Gale, R.J. 1993. Fundamental aspects of extracting species from soils by electrokinetics. *Waste Management* 12: 141–151.

Amira, M., Wilds, C.L., Haltmeier, R.L., Pauling, J.D., and Sontag, J.G., Jr. 1999. Advanced sediment washing for decontamination of New York/New Jersey Harbor dredged materials. In P.E. Randall (Ed.), *Proceedings of Western Dredging Association Nineteenth Technical Conference and Thirty-first A&M Dredging Seminar,* May 15–20, 1999. Louisville Kentucky, Center for Dredging Studies. Texas A&M University, College Station, TX, pp. 567–573.

Baciocchi, R. and Chiavola, A. 2009. Treatment of PAH-contaminated sediments by slurry phase-sequencing batch reactor. *5th International Conference on Remediation of Contaminated Sediments,* Feb. 2–5, Jacksonville, FL.

Bishop, D.F. 1996. Bioremediation of sediments, Seminar Series on Bioremediation of Hazardous Waste Sites: Practical Approach to Implementation. EPA/625/K-96/001. U.S. Environmental Protection Agency, Washington, DC.

Bolam, S.G., Rees, H.L., Somerfield, P., Smith, R., Clarke, K.R., Warwick, R.M, Atkins, M., and Garnacho, E. 2006. Ecological consequences of dredged material disposal in the marine environment: A holistic assessment of activities around the England and Wales coastline. *Marine Poll. Bull.* 52(4): 415–426.

Bray, R.N. (Ed.). 2007. *Environmental Aspects of Dredging.* Taylor and Francis, Leiden, The Netherlands.

Cieniawski, S. 1998. Update from the Great Lakes National Program Office, Remediation Technologies Development Forum, Sediments Remediation Action Team Meeting, Cincinnati, OH, 16–17 September, 1998.

Clarke, A.N., Plumb, P.D., Subramanyan, T.K., and Wilson, D.J. 1991. Soil clean-up by surfactant washing. I. Laboratory results and mathematical modeling. *Sep. Sci. Technol.* 26(3): 301–343.

Clausner, J.E. 1996. *Potential Application of Geosynthetic Fabric Containers for Open Water Placement of Contaminated Dredged Material.* Technical Note EEDP-01-39. U.S. Army Engineer Waterways Experiment Station, Vicksburg, MS.

Colin, D. 2003. Valorisation de sédiments fins de dragage en technique routière. Thèse de l'Université de Caen, France, p. 181.

Comoss, E.J., Kelly, D.A., and Harry, Z. 2002. Innovative erosion control involving the beneficial use of dredge material, indigenous vegetation and landscaping along the Lake Erie Shoreline. *Ecol. Eng.* 19: 203–210.

Dahrazma, B. 2005. Removal of Heavy Metals from Sediment Using Rhamnolipid. Ph.D. thesis, Concordia University, Montreal, Canada.

Dahrazma, B. and Mulligan, C.N. 2007. Investigation of the removal of heavy metals from sediments using rhamnolipid in a continuous flow configuration. *Chemosphere* 69: 705–711.

Davis, A.P., Matange, D., and Shokouhian, M. 1998. Washing of cadmium(II) from a contaminated soil column. *Soil Sediment Contam. Intern. J.* 7(3): 371–393.

DelValls, T.A., Andres, A., Belzunce, M.J., Buceta, J.L., Casado-Martinez, M.C., Castro, R., Riba, I., Viguri, J.R., and Blasco, J. 2004. Chemical and ecotoxicological guidelines for managing disposal of dredged material. *Trends Anal. Chem.* 23(10–11): 819–828.

Dermatas, D., Dutko, P., Balorda-Barone, J., and Moon, D.H. 2002. Geotechnical properties of cement treated dredged sediments to be used as transportation fill. *Proceedings of the 3rd Specialty Conference on Dredging and Dredged Material Disposal,* May 5–8, 2002, American Society of Civil Engineers, Orlando, FL.

Douglas, W.S., Maher, A., and Jafari, F. 2005. Analysis of environmental effects of the use of stabilized dredged material from New York/New Jersey Harbor, USA, for construction of roadway embankments. *Integrated Environ. Assessment Manage.* 1(4): 355–364.

Dragage Verreault. 2008. Technologies décontamination sols et sédiment métaux/hydrocarbures. *Workshop on in situ remediation of contaminated sediment.* Workshop at Concordia University, Montreal, Canada, March 10–11, 2008.

Dubois, V. 2006. Caractérisation physico-mécanique et environnementale des sédiments marins. Application en technique routière. Thesis, Ecole des Mines de Douai, France, p. 311.

El-Shafey, E.I. and Canepa, P. 2003. Remediation of a Cr (VI) contaminated soil: Soil washing followed by Cr (VI) reduction using a sorbent prepared from rice husk. Journal de physique. *IV, International Conference on Heavy Metals in the Environment* No12, Grenoble, France (26/05/2003), vol. 107(1), pp. 415–418.

Environment Canada 1995. Demonstration of a Physico-Chemical Treatment Process for Contaminated Sediment at the Port of Sorel, St. Lawrence Technologies Contaminated Sediment, Ministry of the Environment, Em 1-17/23.

Fenton, H.J.H. 1894. Oxidation of tartaric acid in presence of iron. *J. Chem. Soc.* 65: 899–910.

Fenton, H.J.H. 1895. New formation of glycolic aldehyde. *J.Chem. Soc.* 67. 774.

Fenton, H.J.H. and Jackson, H. 1899. The oxidation of polyhydric alcohols in the presence of iron. *J. Chem. Soc.* 75, 1.

Ferrarese, E., Andreottola, G. and Oprea, I. A. 2008. Remediation of PAH-contaminated sediments by chemical oxidation. *J. Hazard. Mat.* 152: 128–139.

Framework for a Sustainable Dredging Strategy 2007. Final report of Theme 6, Sustainable Dredging Strategies, Report 6.2, 29, http://www.newdelta.org/temp/233314806/NEW!_Delta_Theme_6_report_6-2_-_07_07_04.pdf.

Fujikawa, T., Iguchi, M., Fukue, M., and Sasaki, Y. 2007a. PSFVIP-6: Lifting of small particles up into a cylindrical bath by swirl motion of bubbling jet. *The 6th Pacific Symposium on Flow Visualization and Image Processing*, Hawaii, USA, May 16–19, pp. 117–121.

Fujikawa, T., Sasaki, Y., Fukue, M., and Iguchi, M. 2007b. Ejection of small sedimentary particles into a cylindrical bath by swirl motion of bubbling jet. *J. JSEM* 7(4), 343–348.

Fukue, M., Okusa, S., and Nakamura, T. 1986. Consolidation of sand-clay mixtures. In R.N. Yong and F.C. Townsend (Eds.), *Consolidation of Soils: Testing and Evaluation, ASTM STP892*. American Society for Testing and Materials, Philadelphia, PA, pp. 627–641.

Fukue, M., Yanai, M., Takami, Y., Kuboshima, S., and Yamasaki, S. 2001. Containment, sorption and desorption of heavy metals for dredged sediments. In K. Adachi and M. Fukue (Eds.), *Clay Science for Engineering,* Balkema, Rotterdam, pp. 389–392.

Great Lakes Commission. 2004. Testing and evaluating dredged material for upland beneficial uses. 2nd ed. A Regional Framework for the Great Lakes, September.

Hall, J. 1998. Experimental soft dikes save millions of dollars. 1998 Engineer Update. November 1998, Story 6.

Hazardous Waste Consultant. 1996. Remediating soil and sediment contaminated with heavy metals, Hazardous Waste Consultant, Elsevier Science, November/December, p. 4.10–4.57.

Institute of Gas Technology. 1996. process literature, June.

Interstate Technology and Regulatory Cooperation. 1997. Technical and Regulatory Guidelines for Soil Washing. http://www.itrcweb.org/Documents/MIS-1.pdf.

Jalali, F. and Mulligan, C.N. 2008. Enhanced bioremediation of a petroleum hydrocarbon and heavy metal contaminated soil by stimulation of biosurfactant production. *Geo-Environmental Engineering* 2008, Kyoto, June 10–12.

Jones, K.W., Feng, H., Stern, E.A., Lodge, J., and Clesceri, N.L. 2001. Dredged material decontamination demonstration for the port of New York/New Jersey. *J. Hazard. Mat.* 85: 127–143.

Kahn, Z. and Anjaneyulu, Y. 2006. Bioremediation of contaminated soil and sediment by composting. *Bioremediation J.* 16(2): 109–122.

Karavaiko, G.I., Rossi, G., Agates, A.D., Groudev, S.N., and Avakyan, Z.A. 1988. Biogeotechnology of Metals: Manual. Center for International Projects GKNT Moscow.

Kellar, E.M., Arenberg, E.D., Mickler, J.C., Robinson, L.I., and Smith, S.D. 2009. EN Rx Oxidation process for slow release sediment, soil and groundwater remediation. *5th International Conference on Remediation of Contaminated Sediments*, Feb. 2–5, Jacksonville, FL.

Khodadoust, A.P., Varadhan, A.S., and Bogdan, D. 2009. Enhanced anaerobic biodegradation of PCBs in contaminated sediments using periodic amendments of iron. *5th International Conference on Remediation of Contaminated Sediments*, Feb. 2–5, Jacksonville, FL.

Kuhlman, M.I. and Greenfield, T.M. 1999. Simplified soil washing processes for a variety of soils. *J. Hazard. Mat.* 66: 31–45.

Lee, C.R. and Price, R.A. 2003. Review of phytoreclamation and management approaches for dredged material contaminated with lead. DOER Technical Notes Collection (EDRD TN-DOER-C29) U.S. Army Engineer Research and Development Center, Vicksburg, MS, http://el.erdc.usace.army.mil/dots/doer/.

Löser, C., Andreas, Z., Görsch, K., and Seidel, H. 2006. Remediation of heavy metal polluted sediment in the solid bed: Comparison of abiotic and microbial leaching. *Chemosphere* 65: 9–16.

Löser, C., Andreas, Z., Petra, H. and Seidel, H. 2007. Remediation of heavy metal polluted sediment by suspension and solid-bed leaching: estimate of metal removal efficiency. *Chemosphere* 66: 1699–1705.

Maher, A. 2005. Geotechnical Testing and Beneficial Reuse Evaluation of River Sediments, Palmyra Cove Demonstration Project. Report no FHWA-NJ-2005-014; p. 49. Center for Advanced Infrastructure & Transportation (CAIT), Rutgers, The State University, Piscataway, NJ.

Maher, A., Bennert, T., Jafari, F., Douglas, W.S., and Gucunski, N. 2004. Geotechnical properties of stabilized dredged material from New York–New Jersey Harbor. *Geol. Prop. Earth Mater. 1874*: 86–96.

Maher, A., Douglas, W.S., and Jafari, F. 2006. Field placement and evaluation of stabilized dredged material (SDM) from the New York/New Jersey Harbor, *Marine Geores. Geotechnol.* 24: 251–263.

Massara, H., Mulligan, C.N., and Hadjinicolaou, J. 2007. Effect of rhamnolipids on chromium contaminated soil. *Soil Sediment Contam. Intern. J.* 16: 1–14.

Mechati, F., Roth, E., Renault, V., Risoul, V., Trouvé, G., and Gilot, P. 2004. Pilot scale and theoretical study of thermal remediation of soils. *Environ. Eng. Sci.* 21: 361–370.

Meegoda, J.N. 2008. Feasibility of treating contaminated dredged sediments using ultrasound with acoustic and flow fields. *Workshop on in Situ Treatment of Contaminated Sediments*, Concordia University, Montreal, Canada, Sept. 10, 2008.

Mensinger, M.C. and Roberts, M.J. 2009. Cement-Lock technology demonstration project results with Passaic River sediment. *5th International Conference on Remediation of Contaminated Sediments*, Feb. 2–5, Jacksonville, FL.

Mercier, G., Blais, J.F., and Drogui, P. 2008. Groupe de recherche en assainissement et techologies environmentales, Workshop on in situ treatment of contaminated sediments. Concordia University, Montreal, Canada, March 10–11, 2008.

Merino, J. and Bucalá, V. 2007. Effect of temperature on the release of hexadecane from soil by thermal treatment, *J. Hazard. Mat.* 143: 455–461.

Mulligan, C.N. 2002. *Environmental Biotreatment*. Government Institutes, Rockville, MD.

Mulligan, C.N. and Kamali, M. 2003. Bioleaching of copper and other metals from low-grade oxidized mining ores by *A. niger. J. Chem. Technol. Biotechnol.* 78: 497–503.

Mulligan, C.N., Oghenekevwe, C., Fukue, M., and Shimizu, Y. 2007. Biosurfactant enhanced remediation of a mixed contaminated soil and metal contaminated sediment. *Geoenvironmental Engineering 2007*, Grenoble, France, May 19–24, 2007.

Mulligan, C.N., Yong, R.N., and Gibbs, B.F. 1999. On the use of biosurfactants for the removal of heavy metals from oil-contaminated soil. *Environ. Prog.* 18: 50–54.

Mulligan, C.N., Yong, R.N., and Gibbs, B.F. 2001. Heavy metal removal from sediments by biosurfactants. *J. Hazard. Mat.* 85: 111–125.

Myers, T.E., Bowman, D.W., and Myers, K.F. 2003. Dredged material composting at Milwaukee and Green Bay, WI, confined disposal facilities. DOER Technical Notes Collection (ERDC TN DOER-C33). U.S. Army Engineer Research and Development Center, Vicksburg, MS, http://el.erdc.usace.army.mil/dots/doer/.

Myers, T.E. and Williford, C.W. 2000. Concepts and technologies for bioremediation in Confined Disposal Facilities. DOER Technical Notes Collection (ERDS TN-DOER-C11), U.S. Army Engineer Research and Development Center, Vicksburg, MS, http://el.erdc.usace.army.mil/dots/doer/.

National Research Council. 1997. *Contaminated Sediments in Ports and Waterways. Cleanup Strategies and Technologies*, National Academic Press, Washington, DC.

Nystroem, G.M., Pedersen, A.J., Ottosen, L.M., and Villumsen, A. 2006. The use of desorbing agents in electrodialytic remediation of harbour sediment. *Sci. Total Environ.* 357(1): 25–47.

Olin-Estes, T.J. and Palermo, M.R. 2001. Recovery of dredged material for beneficial use: the future role of physical separation processes. *J. Hazard. Mat.* 85: 39–51.

Palermo, M.R., Francingues, N.R., and Averett, D.E. 2004. Operational characteristics and equipment selection factors for environmental dredging. *J. Dredging Eng.* Western Dredging Association, Vol. 5, No. 4.

Palmer, C.R. 1996. Removal of mercury using the X-Trax thermal desorption system. In *Proceedings of the International Conference on Incineration and Thermal Treatment Technologies*, University of California, Office of Environment, Health and Safety, Irvine, CA.

Papadopoulos, D., Pantazi, C., Savvides, C., Harlambous, K.J., Papadopoulos, A., and Loizidou, M. 1997. A study on heavy metal pollution in marine sediments and their removal from dredged material. *J. Environ. Health* A32(2) 347–360.

Petrovic, M. and Barceló, D. 2004. Seeking harmonisation in assessing sediments and dredged materials, *Trends Analytical Chem.* 23(8): x–xii.

Rankilor, P.R. 1994. The past, present and future for geosynthetics in Indonesia. *Geotextiles Geomembranes* 13(6): 435–456.

Reddy, K.R., Xu, C.Y., and Chinthamreddy, S. 2001. Assessment of electrokinetic removal of heavy metals from soils by sequential extraction analysis. *J. Hazard. Mat.* 84(2–3): 85–109.

Rienks, J. 1998. Comparison of results for chemical and thermal treatment of contaminated dredged sediments. *Water Sci. Technol.* 37: 355–342.

Rittmann, B.E. and McCarty, P.L. 2001. *Environmental Biotechnology: Principles and Applications*, McGraw Hill, New York.

Rodsand, T. and Acar, Y.B. 1995. Electrokinetic extraction of lead from spiked Norwegian marine clay. *Geoenvironment 2000*. American Society of Civil Engineers, pp. 1518–1534.

Sadat Associates Inc. 2001. *Use of Dredged Materials for the Construction of Roadway Embankments*, Vol. I to V. Sadat Associates, Princeton, NJ.

Santiago, R., Inch, R., Jaagumagi, R., and Pelletier, J.-P. 2003. Northern Wood Preservers sediment case study. *2nd International Symposium on Contaminated Sediments*, pp. 297–303.

Seidel, H., Ondruschka, J., Morgenstern, P., and Stottmeister, U. 1998. Bioleaching of heavy metals from contaminated aquatic sediments using indigenous sulfur-oxidizing bacteria: A feasibility study. *Water Sci. Technol.* 37: 387–394.

Selin, H. and VanDeveer, S.D. 2002. Hazardous Substances and the Helsinki and Barcelona Conventions: Origins, Results and Future Challenges, Report presented at the Policy Forum *Management of Toxic Substances in the Marine Environment: Analysis of the Mediterranean and the Baltic*, Javea, Spain, October, 2002.

Semer, R. and. Reddy, K.R. 1996. Evaluation of soil washing process to remove mixed contaminants from a sandy loam. *J. Hazard. Mat.* 45: 45–57.

Severn Estuary Strategy, Joint Issues Summary Report, Environment Agency, London, May 1997, 36 p.

Siham, K., Fabrice, B., Edine, A.N., and Patrick, D. 2008. Marine dredged sediments as new materials resource for road construction. *Waste Manag.* 28: 919–928.

Snell Environmental Group. 1997. Treatability Study Report, Trenton Channel Sediments, Department of Environmental Quality, State of Michigan, Detroit, MI, and USEPA Great Lakes National Program Offices (GL#9985207-01-0), September 1997.

Stern, E.A., Donato, K.R., Ciesceri, N.L., and Jones, K.W. 1997. Integrated sediment decontamination for the NY/NJ Harbor. In *Proceedings of the National Conference on Management and Treatment of Contaminated Sediments*, Cincinnati, OH, 2000, 13–15 May 1997, pp. 71–81.

Tsai, L.J., Yu, K.C., Chen, S.F., and Kung, P.Y. 2003a. Effect of temperature on removal of heavy metals from contaminated river sediments via bioleaching. *Water Res.* 37: 2449–2457.

Tsai, L.J., Yu, K.C., Chen, S.F., Kung, P.Y., Chang, C.Y., and Lin, C.H. 2003b. Partitioning variation of heavy metals in contaminated river sediment via bioleaching: effect of sulfur added to total solids ratio. *Water Res.* 37: 4623–4630.

USACE Detroit District. 1994. Pilot-scale Demonstration of Sediment Washing for the Treatment of Saginaw River Sediments. EPA 905-R94-019. U.S. Army Engineer Detroit District, prepared for the U.S. Environmental Protection Agency, Great Lakes National Program Office, Chicago, IL.

USACE and USEPA. 2003. *Great Lakes Confined Disposal Facilities Report to Congress.* U.S. Army Corps of Engineers–Great Lakes and Ohio River Division and U.S. Environmental Protection Agency–Great Lakes National Program Office, http://www.lrd.usace.army. mil/navigation/glnavigation/cdf.

USEPA. 1993. *Selecting Remediation Techniques for Contaminated Sediments.* Office of Water, EPA-823-B93-001, U.S. Environmental Protection Agency, Washington, DC.

USEPA. 1994. *Solidification/Stabilization of Organics and Inorganics,* Engineering Bulletin, U.S. Environmental Protection Agency, ORD, Cincinnati, OH, EPA/540/S-92/015.

USEPA. 2004. TH agricultural & Nutritional Company Site. RODS Abstract Information, U.S. Environmental Protection Agency, Superfund Information Systems, http://www. epa.gov/superfund.

USEPA. 2005. *Contaminated Sediment Remediation Guidance for Hazardous Waste Sites Report* EPA540R-05-012, U.S. Environmental Protection Agency, Office of Solid Waste and Emergency Response.

USEPA/USACE. 1991. *Evaluation of Dredged Material Proposed for Ocean Disposal.* Report EPA/503/8-91/001. U.S. Environmental Protection Agency and U.S. Army Corps of Engineers, Washington, DC.

Vallée, B. 2008. Restauration environmentale des cellules 1 et 3 des baies du Secteur 103-Zone portuaire de Montréal. *Workshop on in situ treatment of contaminated sediments*, Concordia University, Montreal, Canada, March 10–11, 2008.

Vanthuyne, M., Maes, A., and Cauwenberg, P. 2003. The use of flotation techniques in the remediation of heavy metal contaminated sediments and soils: An overview of controlling factors. *Mineral Engin.*16: 1131–1141.

Virkutyte, J., Sillanpää, M., and Latostenmaa, P. 2002. Electrokinetic soil remediation: critical overview. *Sci. Total Environ.* 289: 97–121.

Wang, S. and Mulligan, C.N. 2004. Surfactant foam technology in remediation of contaminated soil. *Chemosphere* 57: *1079–1089.*

Weyand, T.E., Rose, M.V., and Koshinski, C.J. 1994. Demonstration of thermal treatment technology for mercury contaminated waste, Gas Research Institute, Final Report. June 1994.

Yamasaki, S., Yasui, S., and Fukue, M. 1995. Development of solidification technique for dredged sediments, dredging, remediation, *Containment for Dredged Contaminated Sediments*, ASTM STP1293, pp. 136–144.

Yong, R.N. and Mulligan, C.N. 2002. *Natural Attenuation of Contaminants in Soils.* Lewis Publishers, Boca Raton, FL.

Yong, R.N., Mulligan, C.N., and Fukue, M. 2006. *Geoenvironmental Sustainability*. CRC Press, Taylor & Francis, Boca Raton, FL.

Yozzoa, D.J. and Robert, J.W. 2004. Beneficial use of dredged material for habitat creation, enhancement, and restoration in New York–New Jersey Harbor, *J. Environ. Manag.* 73: 39–52.

Zarull, M.A., Hartig, J.H., and Maynard, L. 1999. Ecological Benefits of Contaminated Sediment Remediation in the Great Lakes Basin, Sediment Priority Action Committee of the International Joint Commission's Water Quality Board, Great Lakes Water Quality Board, http://www.ijc.org/php/publications/html/ecolsed/index.html.

Zentar, R., Dubois, V., and Abriak, N.E. 2008. Mechanical behaviour and environmental impacts of a test road built with marine dredged sediments. *Res. Conserv. Recycl.* 52: 947–954.

Zhao, J.-S., Greer, C.W., Thiboutot, S., Ampleman, G., and Hawari, J. 2004. Biodegradation of the nitramine explosives hexahydro-1,3,5-trinitro-1,3,5-triazine and octahydro-1,3,5,7-tetranitro-1,3,5,7-tetrazocine in cold marine sediment under anaerobic and oligotrophic conditions. *Can. J. Microbiol.* 50: 91–96.

Zhao, J.-S. and Hawari, J. 2008. Biodegradation of nitramines in marine sediment. In Situ Contaminated Sediments Workshop, Concordia University, Montreal, Quebec, Sept. 10.

8 Management and Evaluation of Treatment Alternatives for Sediments

8.1 INTRODUCTION

Selection of the most appropriate remediation technology must coincide with the environmental characteristics of the site and the ongoing fate and transport processes. To be sustainable, the risk at the site must be reduced, and the risk should not be transferred to another site. The treatment must reduce the risk to human health and the environment. Cost-effectiveness and permanent solutions are significant factors in determining the treatment, as is done for Superfund remedies (USEPA, 2005). Development of multiple lines of evidence (LOE) to reach decisions is becoming more frequent and should be utilized by regulatory agencies.

Sites vary substantially, and there can be substantial uncertainty involved in the evaluation process. However, decisions must be made based on the information available. If insufficient information is available to make a decision with some certainty, then the best approach may be to carry out pilot studies of one or more of the treatment techniques. In this chapter, we will examine the means to select the most appropriate technique for site remediation, evaluate the progress of the remediation, and determine the long-term restoration of the site.

8.2 GENERIC FRAMEWORK

The following is a generic framework for sediment assessment from Bridges et al. (2005). It consists of six steps as shown in Figure 8.1. Preassessment is needed to determine the regulatory goals or objectives and to determine how the assessment will be carried out. Sediment quality guidelines (SQGs), international conventions, national and regional laws and regulations, contaminated site assessment, and program guidance are used. Initial assessments help in the identification of minimal- and high-risk sediments. Particular chemicals or toxicity tests might be performed as screening measures. In Canada, a list of chemicals is used for routine use. Sediment assessments can be performed as part of a larger-scale monitoring effort in a watershed. Recently in the United States (USEPA, 2001), the sediments are viewed as having an important role as sources and sinks of contaminants in water bodies of

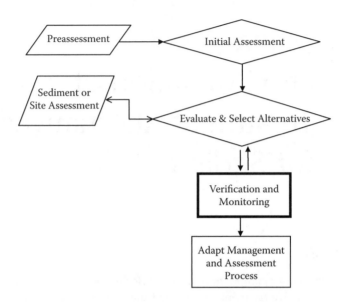

FIGURE 8.1 General assessment/management framework (adapted from Bridges et al., 2005).

restricted use. The contribution of the sediments in an area where there are multiple contaminant sources is a challenge.

The initial assessment includes collecting and analyzing existing data, developing a conceptual model, developing sediment assessment questions, and evaluation of the initial data according to SQGs. Sources of contamination, types of chemicals released from the sources, characteristics of the contaminants, and key receptors should be identified. The conceptual model, as discussed later in this chapter, should describe the sources and the processes (chemical, physical, and biological) going on in the sediment that can lead to exposure by receptors (USEPA, 1998).

8.3   REMEDIATION OBJECTIVES

The concept or aim of remediation of contaminated sediments needs to be properly defined. There are at least two definite objectives in contaminated sediment remediation: (a) reduce the bioavailability of contaminants by removal of the opportunities for bioaccumulation and biomagnification and (b) habitat restoration and regeneration of biodiversity. The various remediation alternatives must be considered and compared carefully. One of the main objectives is to reduce the risk to humans and the environment, as shown in Figure 8.2. This most often is determined by the specific regulatory program governing the site. Risk assessment should compare, evaluate, and rank the risks (USEPA, 1997). Cost-effectiveness is an additional element that must be considered. The risks posed by both the contaminated sediment and the remediation technique must be indentified for the short and long term.

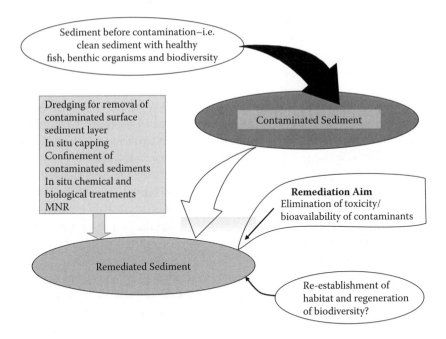

FIGURE 8.2   End points for remediation of contaminated sediments.

The impact on various stakeholders and input from them should be considered. These include local communities and governments and other participants such as those involved in the remediation scheduling, planning, and operation. Communication among the groups is essential. For treatment to be considered, the toxicity or mobility must pose a human risk of 10^{-3} or greater for carcinogens according to the USEPA (1991). Treatment may not always be practical, however. Input from the public can provide useful information regarding the history and current use of the site.

Institutional controls may be needed before treatment can be initiated. Contaminant sources must be identified and controlled before cleanup. Controls can include fishing bans and restricted water use for drinking or swimming, among others. Restrictions on source control may be essential if monitored natural recovery (MNR) or other in situ remediation techniques are to be considered. Restrictions on groundwater flow and use may also be required if the groundwater is a source of the pollutants for the sediment and surface water. Remediation of the groundwater may be required to enhance sediment restoration.

The USEPA identified nine criteria for selecting remediation alternatives. The advantages and limitations of each of the alternatives must be identified, and this is usually done on a qualitative basis. Only some have used a qualitative analysis (Linkov et al., 2004). According to the National Oil and Hazardous Substances Pollution Contingency Plan (NCP) and Comprehensive Environmental Response, Compensation, and Liability Act (CERCLA), cost-effectiveness should be determined based on:

- Effectiveness in the short term
- Effectiveness in the long term
- Capability of the treatment to reduce the toxicity, mobility, and volume of the hazardous materials

The relationship between cost and effectiveness should be determined for the various alternatives, as shown in Figure 8.3. An increase in the complexity of the site often increases the uncertainty of the predicted effectiveness. Often there are tradeoffs between the costs and the effectiveness. The most frequently considered alternatives are MNR, in situ capping, and dredging to remove the sediment from the site. Other approaches include combined techniques and innovative technologies. Impacts and costs must be lower than established technologies. Additional data may be required to prove the effectiveness of the innovative technologies. Many techniques are in the research phase, but should be evaluated and considered. The EPA's Guidance for Conducting Remedial Investigation and Feasibility Studies under CERCLA (USEPA, 1998) should be consulted.

The depth of the contaminated sediment is a major consideration when evaluating bioavailability and site risk. The deeper the sediment, the lower the risk. The potential for bioturbation and erosion that disturb the sediment must be evaluated and be limited. Institutional or engineering controls may be needed. Long-term monitoring is recommended.

As a guideline, the following should be considered during the evaluation of the alternatives. Characteristics of the site, sediment, and contaminants need to be examined. Due to site variability, maximum flexibility should be retained in developing an approach and thus should be site specific.

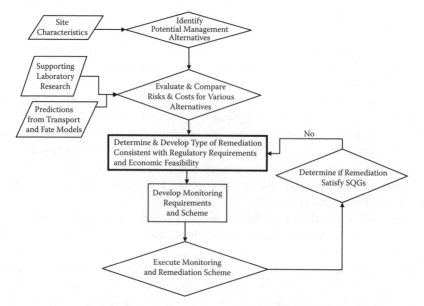

FIGURE 8.3   General activities related to evaluation of management alternatives.

8.4   LINES OF EVIDENCE

Due to the complexity and amount of information required at different sites, multiple lines of evidence are needed to make decisions, since a single line will not be sufficient. Information is biological, chemical, and physical in nature as shown in Figure 8.4. Various stages are carried out to optimize the information required. Decisions can be made at the beginning, but more information may be required to decrease the uncertainty. Various decision points could then be included to determine whether the evaluation should continue or be stopped. Iterations could also be included to allow decisions to be reconsidered or refined.

Principal components in the ecosystem can be used as lines of evidence to address sediment assessment questions. Some endpoints can be sustainable populations of certain fish or birds or biodiversity of species. The LOE can be related to sediment chemistry, toxicity, or benthic community, or organisms in the water column with direct contact with the sediments. Exposure and the effects of exposure on the benthic community are then characterized through lab, field, or modeling assessments.

8.5   EVALUATION OF THE MANAGEMENT ALTERNATIVES

When examining site characteristics, anticipated site uses and structures should be identified. Establishment of new structures such as piers, buried cables, or pilings may interfere with in situ caps or MNR, while older structures can impact dredging activities. Water depths must be sufficient for the establishment of caps so that they do not interfere with navigation, flood control, or boat anchoring. Suitable materials must be available for capping, whereas for dredging, suitable areas for disposal and sediment management are needed.

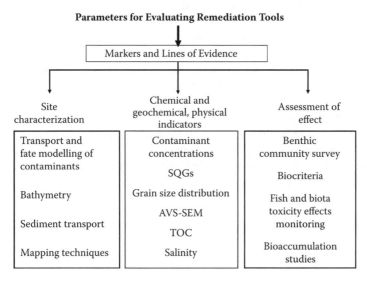

FIGURE 8.4   Lines of evidence indicators.

With regard to the human and ecological environment, exposures to humans during MNR must be low, and rates of recovery must be proceeding at a reasonable rate to reduce toxicity exposures. MNR may be considered in particular where sites are sensitive and potentially damaged by capping or dredging. Capping or dredging can be considered if human exposure is extensive and long term, and thus disruption of the environment is warranted.

The hydrodynamic conditions such as floods or ice scour can influence the effectiveness of capping and thus must be considered during design. MNR, however, is not likely to be affected. The rate of clean sediment deposition is essential for MNR. Groundwater flow into the cap area should be considered to ensure that contaminants are not released to any extent. For dredging, water flow rates should be low to avoid resuspension and dispersal of contaminated sediments downstream.

Sediment characteristics should be determined to predict its behavior and suitability for the proposed remediation. For MNR, the sediment should be cohesive or well armored to minimize resuspension. For capping, high-density/low-water content sediments are more likely to support the cap. For dredging, clean sediment under the contaminated sediment is needed for overdredging. However, overdredging to remove all contaminated sediment must be optimized to minimize sediment handling and disposal. The presence of large amounts of debris in the form of logs, boulders, and metallic waste materials can be detrimental to dredging operations and must be easily removed so that dredging is feasible.

Sediment characteristics are particularly important in considering MNR. Contaminant concentrations in the sediment active zone and biota must be decreasing over time toward risk-based goals. The contaminants must be biodegradable, transformable to less toxic forms, and not able to bioaccumulate substantially. Concentrations should be low over diffuse areas. For both MNR and capping, the fluxes of the contaminants should be low. For dredging, contaminants of high concentrations should be restricted to specific areas to minimize further treatment and disposal costs.

According to the National Contingency Plan (NCP), nine criteria should be considered when evaluating sediment remedial approaches. Tradeoffs must be considered because there are not any zero-risk options. Thus risks and cost must be minimized as much as possible. The criteria include:

- Overall protectiveness
- Compliance with applicable or relevant and appropriate requirements
- Long-term effectiveness and permanence
- Reduction of toxicity, mobility, and volume by the treatment
- Short-term effectiveness
- Implementability
- Cost
- State and community acceptance

Each of the above selection criteria will be considered for MNR, in situ capping, and dredging/excavation. Overall, MNR is based on the use of natural processes to achieve the remediation goals in the short term and possible long term. In situ capping involves the placement of a cap material to protect from the contaminated sediment. The level of protection can be high, but depends on the design, area of contamination, and the maintenance of the cap. By dredging, the contaminated sediment is removed from the site. The level of protection depends on the amount of contaminated sediment that remains at the site. In the event of residual contamination, MNR, capping, or backfilling may be used. Compliance with regulatory requirements for all remedial approaches will depend on the specific contaminant. If fill materials are added for capping, the Clean Water Act or other regulations related to discharge of materials in the water or water obstructions may come into effect. For dredging, disposal regulations for discharges or disposal in landfills (hazardous or municipal) must also be adhered to.

8.5.1 MNR

Long-term effectiveness of MNR depends on the mechanism for isolation or destruction. The overall scheme can be considered as shown in Figure 8.5. Biodegradation of the contaminants will lead to a permanent solution once the levels have decreased. However, if the rate of clean sediment deposition is not sufficient over the long term, bioturbation or other processes that disturb the sediment or diffusion of contaminants may be problematic. In situ capping faces some of the same challenges as MNR. Contaminant migration through the cap via diffusion may occur, or advective flow may be disruptive to the cap. Dredging can provide long-term control if residual contaminant levels are minimal and the disposed contaminated sediment

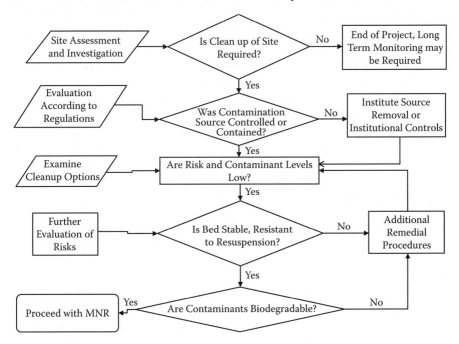

FIGURE 8.5 Evaluation of MNR suitability (from USEPA 2005).

is adequately controlled. Five-year reviews are required for MNR, in situ capping, and on-site disposal facilities. For dredging, these reviews are not generally required once the remediation objectives are met.

MNR is not a treatment, and therefore, reduction of toxicity, mobility, and volume are not achieved through treatment. However, the natural processes can achieve these actions. In situ capping is mainly an isolation process and thus is not a treatment. However, as previously described new types of caps that are reactive are under development. Following excavation, numerous treatment processes as described in Chapter 7 exist. Stabilization/solidification is the most utilized method for reduction of the toxicity of the sediment. Reuse of sediment as agricultural materials, soil replacements, and construction materials has been performed.

In the short term during remediation, impacts can occur to the environment, community, and workers. During MNR, no additional impact to the environment, workers, and community should occur. Due to the continued contact of the benthic community with the contaminated sediments, impact may continue until the remedial objectives are achieved. Fish and shellfish advisories may be advisable at this time. The period of remediation can be highly uncertain and is dependent on the rate of the natural processes and bioavailability of the contaminants.

Since no construction is required, MNR is easily implementable. Uncertainties exist regarding the rates of natural processes and sediment stability. Contingency planning is a necessity if MNR is not successful. The needs for monitoring sediment cleanup can be extensive and costly but are quite established and available. Fish advisories may need to be longer than for other methods. Capital costs are essentially nonexistent.

Finally, regarding acceptance by regulatory authorities and the public, MNR disrupts the site and community the least. However, it is viewed as a "do nothing" technique with long-term potential for contaminant release. Capping is a more active remediation technique than MNR, and the biota can recover faster, but contaminants are still left in place and temporary disruptions occur during placement. Long-term restriction in the use of the waterway can occur. Dredging and excavation removes the contaminants from the site and potentially treats the contaminants. Removal increases the rate of biota recovery and decreases flooding and enhances navigation. Similar to capping, temporary disruptions also can occur during dredging.

8.5.2 DREDGING

The objective of dredging is to remove the contaminated sediment without spreading the contaminated sediment into other areas. An evaluation scheme is shown in Figure 8.6. Dredging can significantly impact the benthic community and bottom habitat for many years. Other impacts are due to resuspension of the sediment during the dredging. Physical barriers such as silt curtains are recommended to limit the area affected by resuspension. The choice of dredging equipment (mechanical versus hydraulic) will also have an influence. Mechanical dredges minimize waste disposal because they handle a higher solids content. Rates of suspension are higher, production rates are higher, and its use is limited to shallow water depths.

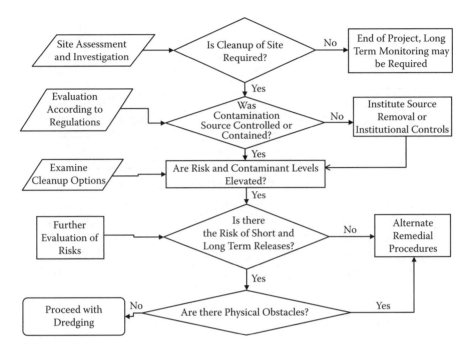

FIGURE 8.6   Evaluation of dredging suitability (from USEPA, 2005).

During dredging, there is the potential for contaminant release that can affect public health. For example, oil can be released, float to the surface, and can volatilize once the sediments are disturbed. Sorbents or other engineering controls may be needed. Increased traffic during transport of dredged materials and water may also have an influence on the local community. Due to the difficulty in estimating residual contamination, the time frame of achieving remediation can be uncertain.

Dredging and excavation are well-developed methods for removal of sediments. Accessibility and the ability to overdredge are required. Geostatistical techniques are being developed to enable better estimation of the location of the contaminated areas. Disposal is often landfills and often requires public and government agency coordination. Disposal in confined aquatic disposal facilities (CADs) are less established and may require extensive monitoring. MNR or capping may also be used if dredging alone is not able to remove all contamination. During dredging, monitoring is required. Environmental dredging equipment is commonly available for most projects. Other equipment for material separation, dewatering, or water treatment may also be required depending on the site. Capital costs are thus high, but monitoring costs should be lower than MNR and capping because it is required mainly during the operation itself.

8.5.3   IN SITU CAPPING

In situ capping is based on the well-established geotechnical principles of slope stability and cap consolidation. It is usually reliable, but monitoring and maintenance of the

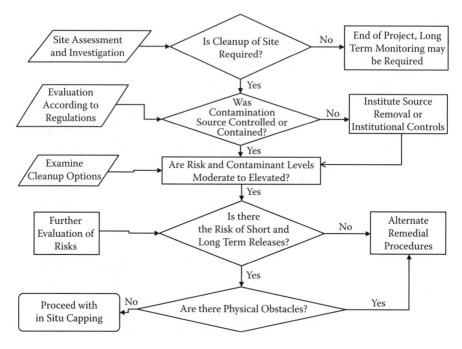

FIGURE 8.7   Evaluation of in situ capping suitability (from USEPA, 2005).

cap are required. A schematic for evaluating the suitability of in situ capping is indicated in Figure 8.7. Monitoring is required until remedial objectives are obtained. The period of operation and maintenance is mainly until the cap is stable and there is little contaminant flux from the cap. Repair of local problems of erosion and disruption are relatively easy. If the cap fails, removal of the sediment can be expensive and difficult. Fish advisories can be required for residual material outside of the cap area. Cap material and equipment for its placement must be suitable and widely available. At some sites, such as urban areas, staging of the cap material can be difficult. Capital, maintenance, and monitoring costs are generally higher that MNR but lower than dredging.

Although some release of contaminants may occur during capping procedures and consolidation, this should be minimal. This may cause some impact on the community and workers. Other impacts may be due to increased traffic during cap placement and transport of materials for the cap. Higher impacts on the environment may occur due to the cap design that encourages recolonization of the benthic community. Once the cap is stable, protection is achieved in a short period of time. However, biota recovery may take several years to achieve.

8.6   SELECTION OF TECHNOLOGIES

According to The Committee on Remediation of PCB-Contaminated Sediments (NRC 2001), all remediation technologies have advantages and disadvantages (Table 8.1), but the overall risk management and reduction should be the most important aspect. Conceptual model development is a vital part of the process and must

TABLE 8.1

Comparison of Dredging, in Situ Capping and MNR Remedial Approaches

Approach	Advantages	Disadvantages
• Dredging	• Restores water depth • Removes contaminants from the site • Lowest uncertainty	• Resuspension, release, and residual contamination possible • Temporary destruction of aquatic community • Costly • Treatment and/or disposal of dredged material is expensive
• In situ capping	• Can be used to reduce exposures quickly • Less material handling, dewatering, treatment, and disposal is needed • Potential for dispersion, volatilization is lower • Conventional and available materials and equipment used • Less disruptive of communities and can enhance the habitat • Can be cost-effective	• Contaminated sediment is left in place • Cap may be damaged or dispersed by storms/floods • Institutional controls are often necessary • Considered an emerging technology • Water depth is reduced • Long-term monitoring and maintenance needed
• MNR	• No disruption of the water column or aquatic communities • Low cost	• Contaminants remain in place • Processes can be slow • Storms/floods can affect the area • Institutional controls are often necessary • Long-term monitoring required • Public acceptance is not easy to obtain

be site specific. Therefore, human and ecological exposure to contaminants must be reduced, and risks of the technology implementation during remediation must be low. Reduction of the bioavailability or bioaccessibility of the contaminants can be achieved by reducing the exposure potential. This can be accomplished by burial or isolation of the contaminants, destruction of the contaminants, or removal from the site. These questions are addressed for LOEs and assessment tools.

For MNR, exposures to contaminants by the organisms at the sediment surface and within the sediment can continue until the contaminants are degraded. Sediment stability is another issue for long-term risks. Low sedimentation rates and other changes can reduce the rate of MNR. Erosion or human disturbance (such as boat wash) may disperse the contaminants.

For in situ capping, contaminant releases may continue until the capping is completed. This exposes workers, organisms and the community to the contaminants.

The contaminants outside the main contaminant area may be left untouched and subject to disturbance. The placement of the cap is disruptive to the public and the benthic community. For dredging and excavation, many of the risks are similar to capping. Additional risks occur, due to sediment handling, to the workers and to the community for subsequent disposal if no treatment is performed. Further risks are associated with disposal such as generation of side-streams during treatment.

The decision for no action may be appropriate if there is no current or potential threat to the environment and the public. If fish bans or consumption advisories are required, then it is obvious that there is a risk, and no action would not be appropriate. Monitoring over the long term may be necessary if there is uncertainty regarding the risk to the public. The difference in monitoring for MNR is that the objective of MNR is to prove that the contaminant levels are reducing.

Therefore, a net overall environmental benefit should be obtained. Short- and long-term environmental effects need to be balanced. Solutions must be specific for the site. Combined approaches may be more appropriate than one alone, particularly for large sites as shown in Figure 8.8. Dredging (for high-level contamination) can be combined with thin capping (for medium contamination) and then MNR (for low level) if not all of the material can be removed.

8.7   MANAGEMENT PLAN

To determine the best approach for a site, a conceptual model should be developed for evaluation of the various approaches and risk identification (Figure 8.9). Both

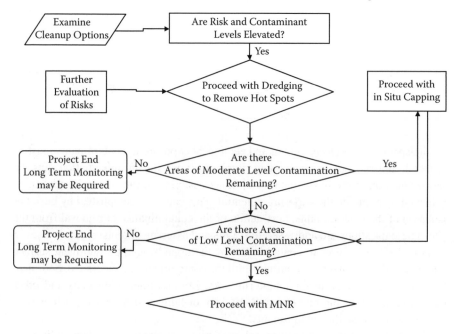

FIGURE 8.8   Combined approach for remediation of contaminated sediments.

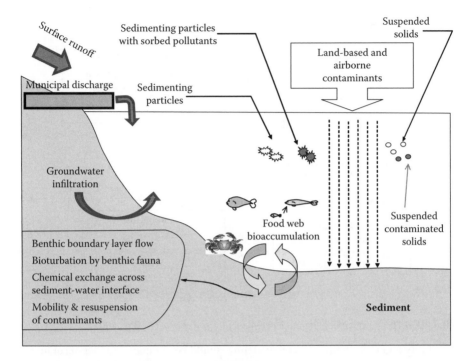

FIGURE 8.9   Conceptual model of contaminant sources and risks.

short-term and long-term objectives must be fulfilled. Source control is highly desir-able (USEPA, 2002) because a continuous input of contaminants can defeat remedia-tion effectiveness and recontaminate the area. A combination of approaches may be appropriate if the site is highly heterogeneous. Estimates of residuals and contami-nant releases from in situ and ex situ approaches must be realistic to enable compari-sons. Sediment stability and contaminant transport are issues of concern. Models for evaluation of these phenomena are used to determine these differences and should incorporate the transport pathways (Figure 8.10).

Monitoring during the remediation and after is highly important and thus should be planned and initiated at contaminated sites. It can also be used to document short- and long-term impact and compare to impacts predicted in an Environmental Impact Assessment (EIA). Monitoring will indicate what the risk is during the remediation, if the cleanup objectives have been achieved, and the permanence of the remediation, if recontamination is a possibility. If a cap, then it must be stable and not subject to erosion. Recovery of biota including fish and benthic organisms is essential both on the short and long term. Five-year assessments are often required to determine the long-term effectiveness of the remediation. Uncertainties can also be reduced, and experience can be gathered for future projects by monitoring.

Chemical, physical, and biological data should be collected at sufficient times and places. Baseline data is usually collected during the initial site characterization phase. The nature and extent of contamination is determined to evaluate remediation

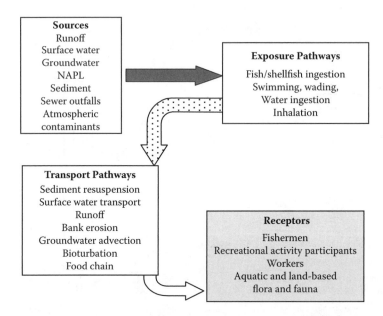

FIGURE 8.10   Conceptual site model elements for sediment (USEPA, 2005).

feasibility and risk. Predictions by modeling should be compared to long-term monitoring data, and thus the monitoring plan should be appropriate. Background data is
required from related uncontaminated areas to enable establishment of background
levels and also to determine changes to indicate sources of contamination such as
runoffs.

Monitoring at sediment sites can be complicated due to multiple media including surface water, biota, groundwater (if applicable), floodplain soils, and of course,
sediment. Sources of contamination can be diverse, and thus a wide variety of contaminants often is found. Sites can also be very large and heterogeneous. Goals for
remediation are risk based, and the relationship between sediment, contaminants,
and biota is complex. The frequency and extent of the monitoring should be established from the beginning. Electronic databases may be desirable for sharing results.
Statistical or other quantitative methods for analysis of the data should be identified.

Six steps have been developed as a monitoring plan by the EPA's Monitoring
Guidance (USEPA, 2004). They include:

- Identification of the monitoring plan objectives
- Development of monitoring plan hypotheses
- Formulation of monitoring decisions guidelines
- Design of the monitoring plan
- Conducting of the analyses and characterization
- Establishment of the management decisions

To develop a monitoring plan, the objectives must be clear and specific, particularly due to the limited funding for monitoring. Useful data needs to be collected and

often in consultation with the public, local agencies, and other parties. Often funding agencies such as the World Bank require Environmental Management Plans which are integrated into EIA before permits can be issued. Physical, chemical, and biological objectives should be identified. However, biological endpoints may be more difficult. Toxicity or bioassessment tests may be better for examination of the effects on organisms. In the short term, acute toxicity tests on an organism or measurement of a specific contaminant such as PCBs may be used, whereas for the long term, species diversity or measurements in fish tissue of a contaminant may be more appropriate. Usually a combination of physical, chemical, and biological approaches is needed to determine risk reduction.

To develop a monitoring plan, hypothesis statements related to the remediation or outcome can be formulated. For example, has the remediation enabled the goals and objectives of the remediation to be achieved? A flow chart such as shown in Figure 8.11 can be used to identify the site activities and potential outcomes. Some hypotheses can be: Has the PCB concentration in the sediment reached the goal of 0.5 ppm, or has the PCB concentration in the fish tissue been reduced to 0.05 ppm?

Various types of monitoring plans are used (Bray, 2008): surveillance monitoring, feedback (Europe), or adaptive monitoring (United States) and compliance monitoring. Surveillance monitoring is performed to verify the hypotheses put forth during the preparation of the project. Usually the national or local authorities are responsible for this type of monitoring. Some other hypotheses are related to the background conditions, such as contamination levels, or conditions at the site, or

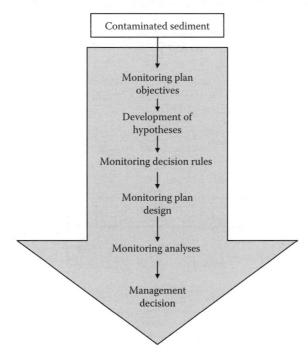

FIGURE 8.11   Simple protocol for monitoring of a contaminated site.

particular operational parameters to be used. International conventions such as the London Convention or local conventions must be consulted before a monitoring scheme is devised.

Feedback monitoring (Figure 8.12) is performed to evaluate if any environmental criteria are exceeded in a timely fashion to avoid down time. It is the responsibility of the project owner; it is comprehensive and thus costly to carry out. It is used where environmental criteria are strict or the impacts must be strictly observed. Compliance monitoring is performed to ensure that contractual restrictions are executed by the contractor. For example, there may be restrictions regarding the breeding seasons, the location and transport of the sediments, dredging depths, etc. The requirements must be clear to avoid future problems. In the case of noncompliance, corrective measures would be necessary.

Once the goals are established, the rules should be established to enable the choice of the various actions. The contaminants of interest, the remediation action, the goals of the remediation, and various alternatives to achieve the goal should all be included. The time frame should also be considered, because a particular action may not be proceeding quickly enough (such as natural recovery), and thus an alternative must be employed (such as in situ capping).

To design the monitoring plan, the frequency and location needs to be identified. Locations to obtain baseline data should be identified. The site conditions will influence the location of the sampling points as transport of the contaminants will be downstream or through the food chain. Statistical approaches can be used to establish data trends to be able to draw conclusions. A balance is needed between

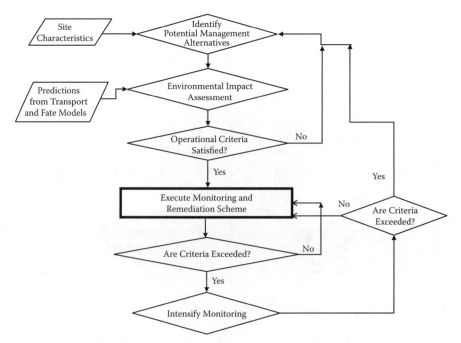

FIGURE 8.12 Feedback monitoring scheme (adapted from Bray, 2008).

fewer samples which cost less and the required data to design an appropriate plan. Appropriate indicators may also be used.

Collection and analysis of the data is an important step in the monitoring plan. It allows the project manager to determine if the remediation is proceeding in the right direction. Decisions at this point regarding the continuation or modification of the monitoring or remedial activity can be taken. For example, if the contaminant concentration in the fish tissue has been decreasing significantly over the first two years but has not reached the five-year goal, then monitoring would continue. Finally, various contingency plans should be made in advance. If the data shows that the remediation is proceeding well, the frequency of monitoring could decrease. However, to obtain more data, more frequent sampling or sampling in more locations may be required.

Methods for analysis and sampling methods have been described in Chapter 4. These are continually being improved. Modeling can complement the monitoring to evaluate local variations of stochastic nature. If no modeling is performed, then the affected area needs to be completely monitored. Physical measurements are needed to evaluate erosion and/or deposition of sediments, depth of the sea or riverbed, suspended solid concentration, groundwater flow, particle size, water flow rates, and sediment homo- or heterogeneity. Acoustic Doppler current profilers (ADCP), satellite images, and aerial photography are now being developed to provide indications of suspended sediment content (Bray, 2008). Chemical data is needed for information on the sediment chemistry, and to evaluate biodegradation and partitioning of the contaminant and total organic carbon content. Biological testing can include tests for toxicity, bioaccumulation, biodiversity, and food chain effects. Sonar and radar are useful for studying fish and bird behavior, respectively.

Monitoring approaches differ according to the remedial action. Monitoring of the source control is highly important, as seen in Figure 8.13. For natural recovery, monitoring is a key issue. The natural processes include sedimentation, accumulation, and sediment and contaminant transport. Contaminant levels and degradation products in the sediment, surface water, and biota must be monitored. Recovery of the biota can be determined through sediment toxicity, benthic community characterization, and diversity. Extensive monitoring can be required initially while recovery is occurring and if there is a disturbance such as a storm. If modeling has been used, monitoring can be used to validate the model. When cleanup is obtained, then monitoring may still be needed to ensure that the remediation is still effective, particularly if burial was the major mechanism.

For in situ capping, monitoring of the cap itself is required to ensure that it is placed and performing properly. This can be performed through bathymetric surveys to determine the cap thickness, sediment cores, sediment profiling cameras, and chemical resuspension monitoring. Diver observations or viewing tubes are also employed in shallow waters. Storms, ice scour, and other events including boat wash can disrupt the cap. Repair of the cap may be necessary.

Long-term monitoring is needed to ensure that the cap has not eroded. Monitoring of the biota to ensure recovery is required. Recolonization of the cap area and bioturbation of the cap may be monitored. Excessive bioturbation may require the supplementation of the cap material. Contaminant bioaccumulation in fish and benthic

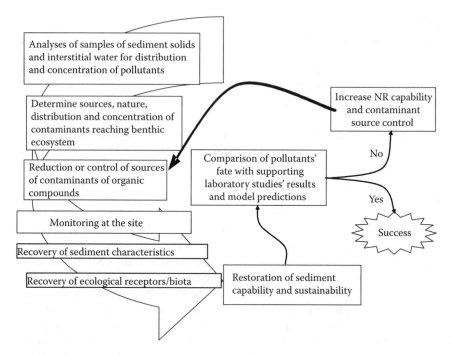

FIGURE 8.13 Required activities for sustainability assessment and monitoring of remediated sediment based on distribution and concentration of contaminants in remediated sediment.

organisms can also be determined. Chemical fluxes should be determined to evaluate for potential influxes from groundwater recharge, or from the sediment to the surface water. Both capped and noncapped areas should be monitored.

For dredging, monitoring is required during the dredging, transport of sediments, dewatering, and disposal and/or treatment. Residual sediment needs to be monitored for benthic recovery, bioaccumulation, and contaminant concentrations in the sediment to determine if remediation objectives have been met. Further dredging or backfilling may be required if the objectives are not met or remain being met. Surface water during dredging should be monitored for suspended solids, total and dissolved contaminant concentrations, and fish toxicity or caged mussel intake. Pilot tests may be needed prior to dredging to evaluate the potential effect on surface water. Containment barriers such as silt curtains can be installed during the dredging. Water from dewatering of sediments needs to be monitored. Air monitoring is also needed during dredging/excavation, disposal, and treatment. This is particularly important in the case of volatile contaminants. Short-term adverse effects of dredging include accumulation of contaminants in the tissues of caged fish and other biota. There is a lack of data at many sites. Long-term monitoring is not performed over an extensive period of time. Capping or backfilling is sometimes necessary if residual concentrations are significant (NRC, 2007).

Prior to disposal of residuals, the following scheme should be undertaken (Figure 8.14). Beneficial use should be evaluated based on economics and feasibility.

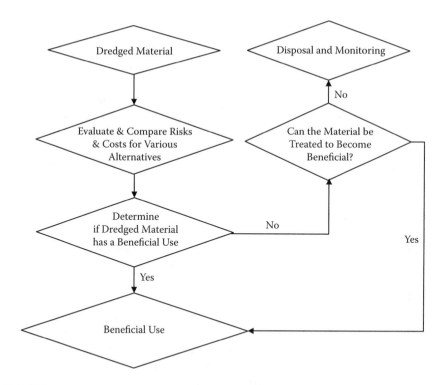

FIGURE 8.14 Enhancing the sustainability of dredged or removed contaminated sediments.

If this is not possible, then disposal will have to be performed. Treatment of residuals at a disposal facility will also need to be evaluated to ensure contaminants are not released. Toxicity characteristic leaching procedure (TCLP) tests can be used on the residuals to determine if there is the potential for contaminant leaching. The facility itself needs to be monitored to ensure that the structure, leachate collection systems, liners, and treatment facilities are not compromised. Storage capacity must not be exceeded, and consolidation/compactions, gas production, and organic matter decomposition must be monitored. Leakage of the contaminants from the CDF or landfill to groundwater or surface water needs to be identified through monitoring. Runoff from the facility also needs to be evaluated. Noise, dust, and odor are other considerations. If there is a cap in the disposal area, revegetation by plants or recolonization by animals needs to be monitored.

Overall, monitoring must be well planned and executed to ensure that the requirements are met. Sampling, analysis, quality control, and reporting must be performed professionally. Often analytical laboratories are used to ensure the procedures are timely and at the lowest cost. The benefit of the remediation work must outweigh any impact caused. Therefore, advantages versus disadvantages must be compared, in particular, to ensure that the remediation method is sustainable and long-term risk reduction has occurred.

8.8   SUSTAINABLE REMEDIATION

When treated successfully, sediments can be considered without risk to humans and the environment. However, other aspects must also be considered, such as habitat restoration, species preservation, and biodiversity regeneration, if the remediated sediment will be maintained over a long period of time. By definition, sustainability means the ability to sustain, maintain, or preserve the ability of the system to maintain the initial uncontaminated condition or state. Sustainability of remediated sediments is thus the ability of the remediated sediments to be restored and maintained in remediated state. The key to sustainability assessment is identification of the basic objective of remediation. As previously discussed, the intent of remediation is to minimize and/or eliminate health risk to humans. The remedial solutions must ensure that contaminants in the sediment must not resuspend or resolubilize so that they either bioaccumulate or become bioavailable.

For a remediated sediment treatment to become sustainable, the sediment must (a) not require retreatment to preserve its remediated state and (b) reestablish its original uncontaminated benthic ecosystem. Given the various sources and inputs of contaminants and the various natural processes in the benthic zone, sustainable remediated sediment preservation may not be easily achieved. The fundamental problem is that strategies and technologies must be developed for the preservation of sustainable remediated sediment. Maintaining the state of remediated sediments will avoid threats to human health. Retreatment of contaminated remediated sediments is costly and needs to be avoided and may also pose further threats to the ecosystem.

A sustainability assessment of remediated sediments is a procedure that is designed to determine if the remediated state of the sediment can be preserved. The results of the assessment can evaluate if recontamination of the remediated sediment can occur, if measures are needed to enhance the remediation, or if further remediation of the recontaminated sediment is required. If habitat restoration and species preservation or biodiversity regeneration are the final objectives, at least four interacting components must be acquired and include:

- The characteristics of the sediments
- The characteristics of the contaminants
- The treatments used
- The requirements for sustainability

Resuspension and remobilization of contaminants of the contaminated sediments must be avoided. The information obtained will also allow one to determine the best or most technically and cost-effective means for treatment.

The various strategies, as previously discussed (Figure 8.15), for remediation of contaminated sediments provide for different results concerning how the contaminants in the sediments are neutralized or eliminated. The nature of the remediated sediment will have a direct influence on the strategies and capabilities for sustainability of the remediated sediment to be achieved. The requirements for remediated

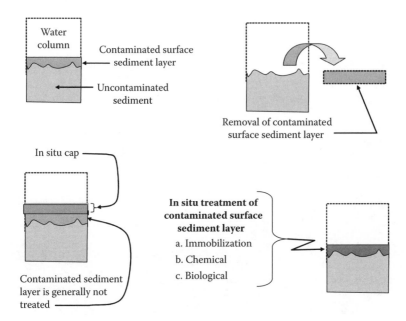

FIGURE 8.15   Various remediation procedures and strategies.

sediment sustainability assessment will include short- and long-term human health risks, regulatory attitudes and goals, economics, and site-specific parameters.

Unless further actions are undertaken, the problems caused by human activities will continue and may escalate. The main challenges are related to land management, energy utilization, consumption of resources, waste management, reduction of pollution, and water resource management. Evaluation of the alternatives should include a life cycle assessment of the remediation alternative. Some of these factors are indicated in Figure 8.16. To do this the overall process must be considered, including the assessment, remedial process, disposal of materials (if applicable), and monitoring. Energy consumption, emissions, material use, and wastes are major components of this assessment.

8.9   STRATEGY FOR REMEDIATED SEDIMENT SUSTAINABILITY

Remediated sediments must not pose indirect or direct threats to human health. The sustainability goal for remediated sediment is the long-term preservation of its remediated state. For this to occur, it is necessary that the treatment technology used in remediation be effective, and that the inputs of contaminants into the ecosystem (including suspended solids, noxious airborne gases, and particulates) are eliminated or substantially eliminated.

Figure 8.17 indicates a five-part strategy that can be implemented to obtain sustainability of the remediated sediment. To reduce the suspended solids above the sediment, cleanup of the water would be required, as discussed in Section 6.2. If the source of contaminants is controlled, then this would be a one-time event. The

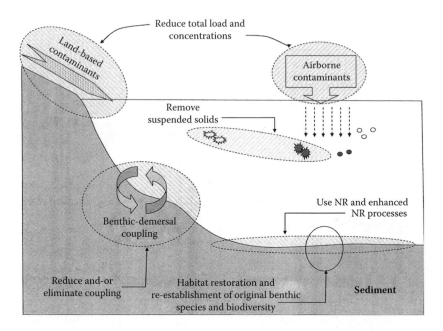

FIG 8.16   Schematic illustration of the five-part strategy to obtain complete sustainability of the remediated sediment.

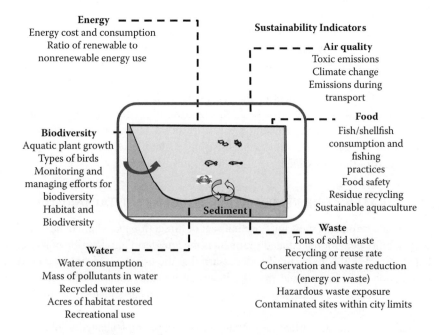

FIGURE 8.17   Required activities for sustainability assessment and monitoring of remediated sediment based on distribution and concentration of contaminants in remediated sediment.

suspension of the solids may also be due to the treatment action itself, such as dredging of hot spots or storm events or extensive bioturbation that breaches an in situ cap. In areas that are not technically or economically feasible to remediate, natural recovery processes can be exploited or enhanced, if necessary. These natural processes, which have been previously discussed, must be able to neutralize, decontaminate, and/or reduce the bioavailability of contaminants for the benthic animals. Finally, ultimate remediation encompasses the restoration of habitat and reestablishment of benthic species and biodiversity similar to the precontamination state of the zone. Monitoring must play an essential role in ensuring the short- and long-term remediation of the site.

Sustainability indicators used for monitoring the remediated state of contaminated sediments can be very simple or complex. The choice of indicators to establish sustainability could include (a) level of bioturbation and bioirrigation, (b) distribution of partitioned contaminants, (c) nature and concentration of contaminants in the interstitial water, (d) biological diversity, etc. Figure 8.18 provides a protocol leading to the sustainability assessment of a remediated sediment based on the distribution and concentrations in the remediated sediment.

Several criteria can be employed to declare when sustainability of the remediated sediment has been achieved. As shown in Figure 8.18, the presence of contaminants or SQGs in the remediated sediment will be frequently evaluated to determine sustainability. The indicators may also refer to the distributions and concentrations of

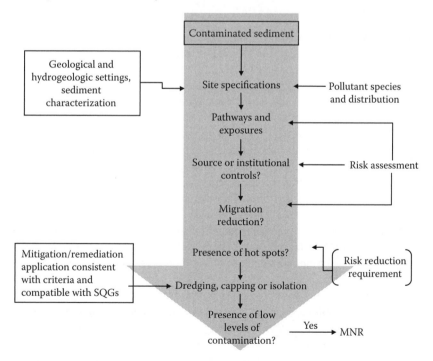

FIGURE 8.18   Simple protocol for rehabilitation of a contaminated site.

target species of contaminants in the surface sediment layer at the top of an in situ cap, the MNR layer, or the layers treated by other remediation processes.

To determine the nature and distribution of contaminants attached to sediment solids and in the pore water, laboratory tests and studies for partitioning of the kinds of contaminants found in the sediment are needed along with studies on the intermediate products of the organic chemical pollutants found in the sediment. Potential resuspension and remobilization of contaminants in the turbulent layer must be predicted. To accomplish this, fate and transport models must be developed and implemented to predict the distribution of contaminants found in the remediated sediment over time. The required activities can be considerable, although many of the detailed activities are unnecessary, but the risk to human health and the ecosystem must be minimized. The risk of recontamination of the remediated sediment is a central issue in the sustainability assessment, and thus the control of the input of contaminants is vital.

Remediation of contaminated sediments entails contaminant removal, isolation, or reduction of the toxicity. Remediation goals are mainly focused on the contaminant levels. Restoration of the habitat and biodiversity is gaining more attention, because this is the ultimate goal for the benthic ecosystem. Specification of the indicators requires baseline information. Species diversity, natural communities, and other related biomarkers need to be included as sustainability indicators.

8.10   CONCLUDING REMARKS

Remediated sediments do not necessarily mean that contaminated sediments have been remediated to the extent that all the contaminants in the sediments have been removed, or that the sediments are devoid of contaminants. Based on site-specific characteristics, the most appropriate remediation technology can be chosen. The important lesson to be learned is that remediated sediments mean that the threats posed by the contaminants in the sediments have been neutralized or eliminated. Except for physical removal of all contaminated sediment layers (which even in the case of dredging is often not obtained), there will be contaminants remaining, in one form or another, in the sediments remediated with currently used technologies. Successful application of treatments involves reducing the bioaccessibility and bioavailability, in addition to the resuspension and remobilization of the contaminants.

Sustainable remediation practices require (a) source control of contaminants entering the ecosystem, (b) utilization of the natural processes in the ecosystem to maintain the remediated state of the sediment, and (c) restoration of habitat and reestablishment of biodiversity. Contaminants accumulate both in the sediment and the food chain. Global warming will put further stresses on the fresh- and saltwater environments. Eutrophication in addition to accumulation of contaminants is also a major concern. Figure 8.19 shows an illustration of a simple strategy for rehabilitation of the aquatic geoenvironment. Nutrient levels must be balanced so that limitations and excesses are avoided. To avoid obtaining a sterile sediment bed, a natural purification system needs to be established so that habitat restoration can occur. Human intervention in providing

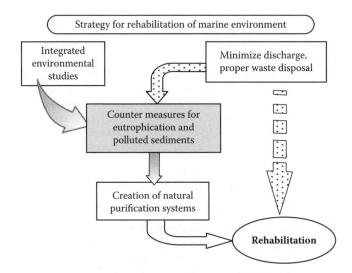

FIGURE 8.19 A strategy for rehabilitation of the marine environment.

the necessary elements for restoration of habitat and reestablishment of biodiversity, after or during remediation of the contaminated sediment, will provide for sustainable remediated sediment preservation.

REFERENCES

Bray, R.N. (Ed.). 2008. *Environmental Aspects of Dredging.* Taylor and Francis, Leiden, The Netherlands.

Bridges, T.S., Berry, W.J., Della Sala, S., Dorn, P.B., Ells, S.J., Gries, T.H., Ireland, D.S., Maher, E.M., Menzie, C.A., Porebski, L.M., and Stronkhorst, J. 2005. A framework for assessing and managing risks from contaminated sediments, In R.J. Wenning, G.E. Batley, C.G. Ingersoll, and D.W. Moore (Eds.), *Use of Sediment Quality Guidelines and Related Tools for the Assessment of Contaminated Sediments.* Society of Environmental Toxicology and Chemistry, Pensacola, FL, pp. 227–309.

Linkov, I. Varghese, A., Jamil, S., Kiker, G.A., Bridges, T., and Seager, T. 2004. Multi-criteria decision analysis: framework for applications in remedial planning for contaminated sites. In I. Linkov and A. Ramadan (Eds.), *Comparative Risk Assessment and Environmental Decision Making.* Kluwer, Amsterdam.

NRC. 2001. *A Risk-Management Strategy for PCB-Contaminated Sediments. Committee on Remediation of PCB-Contaminated Sediments.* Board on Environmental Studies and Toxicology, Division on Life and Earth Studies, National Research Council, National Academies Press, Washington, DC.

NRC. 2007. *Sediment Dredging at Superfund Megasites: Assessing the Effectiveness.* National Research Council, National Academies Press, Washington, DC, 294 pp.

USEPA. 1991. Handbook: Remediation of Contaminated Sediments. EPA 625/91/028. United States Environmental Protection Agency, Office of Research and Development, Cincinnati, OH.

USEPA. 1997. Ecological Risk Assessment Guidance for Superfund: Process for Designing and Conducting Ecological Risk Assessment. Interim Final. U.S. Environmental Protection Agency. Office of Solid Waste and Emergency Response, Washington, DC, EPA 540/R-97/006/.

USEPA. 1998. EPA's Contaminated Sediment Management Strategy. U.S. Environmental Protection Agency. Office of Water, Washington, DC EPA 823/R-98-001. The strategy and a fact sheet on this document at http://www.epa.gov/waterscience/cs/stratndx.html.

USEPA. 2001. Forum on Managing Contaminated Sediments at Hazardous Waste Sites. U.S. Environmental Protection Agency. Office of Emergency and Remedial Response, Washington, DC. Proceedings available at http://www.epa.gov/superfund/health/con-media/sediment/meetings.htm.

USEPA. 2002. Principles for Managing Contaminated Sediment Risks at Hazardous Waste Site, U.S. Environmental Protection Agency, Office of Emergency and Remedial Response, Washington, DC, OSWER Directive 9285.6-08.

USEPA. 2004. Guidance for Monitoring at Hazardous Waste Sites: Framework for Monitoring Plan Development and Implementation OSWER Directive 93255.4-28.

USEPA. 2005. Contaminated Sediment Remediation Guidance for Hazardous Waste Sites. Environmental Protection Agency, Office of Solid Wastes and Emergency Response, EPA-540-R-05-012, OSWER 9355.0-85, Washington, DC.

9 Current State and Future Directions

9.1 INTRODUCTION

Estimates in the amount of contaminated sediment vary widely with a large uncertainty. There will be continual requirements for dredging to maintain adequate depths in waterways, particularly in ports and harbors. Often contaminated sediment is part of the dredged material. In addition, there are many inland surface waters that become contaminated due to agricultural, municipal, and industrial discharges. The Great Lakes are a notable example. A lack of dredging impacts harbor management and waterfront development. In addition, the contaminants pose risks to human health and the ecology in the surface water. Cost-effective sediment remediation technologies are required due to their potential impact on infrastructure renewal and commerce due to delays in adequate sediment remediation strategies.

In this book, two main approaches have been examined, in situ and ex situ. Environmental dredging requires evaluation of the risks of dredging, determination of disposal methods, and/or potential beneficial use. Depending on site conditions, in situ management may be preferable and may pose less risk to human health, fisheries, and the environment. Both short-term and long-term risks must be evaluated for the in situ and ex situ options. This approach is summarized in Figure 9.1. A management option where all aspects coincide will be the most appropriate.

A life cycle approach is needed where both short- and long-term aspects must be examined on an environmental, social, and economic basis. To work toward sustainability, waste must be minimized, natural resources must be conserved, landfill deposition should be minimized, benthic habitats and wetlands must not be lost and must be protected. Innovative integrated decontamination technologies must be utilized.

We will examine, also, where developments are needed. Since the 1990s, removal and treatment were the most common methods of dealing with contaminated sediment. Most R&D in North America and Europe thus focused on method development. However, removal of sediments and subsequent management of the dredged materials can increase risk to human health and cause ecological damage without substantial benefit (Thibodeaux et al., 1999). We have also seen that leaving the sediments in place without engineered remediation (monitored natural recovery, MNR) may also be more appropriate. However, the fate and transport of contaminants must be understood more thoroughly to develop appropriate strategies.

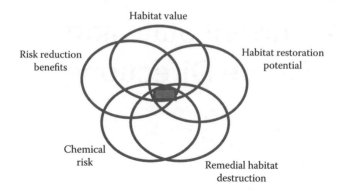

FIGURE 9.1   Decision basis for cleanup and restoration (adapted from Ludwig et al., 2006).

9.2   DISPOSAL AT SEA

Disposal at sea is governed by international conventions. However, various countries use different approaches for disposal of dredged materials at sea. Some use physical/chemical characteristics of the sediment, while others use the contamination levels and other ecotoxicological effects before disposal. Coastlines may also be defined differently. Sampling and testing of the samples are expensive and thus difficult to implement. For disposal on land, international conventions do not apply but may be beneficial to provide a degree of standardization.

Canada has more than 243,000 km of coastline (the longest of any country) (Parks Canada, www.pc.gc.ca/progs/amn-nmca/intro_e.asp). Although most is uncontaminated, there exist problems in harbors, estuaries, and areas near the shore. Disposal of dredged materials is permitted offshore. Only recently, contaminated dredged materials could be disposed of in marine and estuarine waters. Chemical, biological, and physical monitoring is an essential part of the Environment Canada Disposal at Sea Program.

Lower action levels are based on cadmium (0.6 mg/kg), mercury (0.75 mg/kg), total polycyclic chlorinated biphenyls (PCBs) (0.1 mg/kg), and total polyaromatic hydrocarbons (PAHs) (2.5 mg/kg). If levels are above these levels, further action is required such as further analysis based on biological measurement. If elevated biological or contamination levels are found, then further monitoring or determinations of sediment quality are required. A recent Contaminated Dredged Material Management Decisions Workshop (CDMMD) (Jim Osborne Consulting, 2006) recommended that additional contaminants be added, including a full metal scan. A toxicity test should also be added. To ensure consistency throughout Canada, cooperative agreements should be developed with the Great Lakes program and the St. Lawrence Plan for Sustainable Development. The Great Lakes program has focused on containment and is proposing a risk assessment process. In the St. Lawrence River, dredging and disposal of sediments have been approved in the section of the St. Lawrence not covered by the Canadian Environmental Protection Act (Environment Canada, 1999). They have developed a risk assessment approach. Permits have not been issued for disposal of contaminated sediments because the bioassay option is not certain for obtaining a permit, and thus most contaminated sediments are left in place or disposed of on land.

Biological assessment is based on the following test methods: an acute toxicity test for marine or estuarine amphipods, a fertilization assay with echinoids, a solid-phase toxicity test with photoluminescent bacteria, or a bedded sediment bioaccumulation test with bivalves. Further monitoring, site closure, or remediation should be considered if the substances are in excess of the screening level and the acute test or two or more tests are failed. Monitoring is performed at sites greater than 100,000 m^3 every 5 years at least.

As other countries such as Japan will discontinue sea disposal, this policy seems to be going in the opposite direction. Barring in-water disposal of sediments, however, may limit the beneficial use of the sediment. This is because enhancement of the habitat by strategic placement of the sediment would be prohibited. This aspect, habitat restoration or enhancement, should be integrated into sediment management plans.

9.3 BENEFICIAL USE OF DREDGED MATERIALS

Sediments are a resource and thus should be used as such. Dredging moves material from one place to another. The categories of beneficial use of dredged materials include (USACE, 2007):

- Development of wetlands, islands, and other habitats for various birds
- Nourishment of beaches
- Aquaculture
- Recreational areas such as parks
- Agriculture, forestry, and horticulture
- Reclamation of strip mines and for solid waste management
- Erosion control and shoreline stabilization
- Construction and industrial uses for ports, airports, and other uses
- Material transfer for fill, roads, parking lots, dikes, levees
- Other purposes

Managing the sediment involves choosing the most appropriate option. Uncontaminated dredged material has been used frequently for many uses. However, more extensive use of low to moderately contaminated sediment is desirable. The environmental effects (including social, economic, and political effects) of each must be evaluated. Advantages and disadvantages must be compared. No standard integrated procedures exist for evaluation of a project. The USEPA/USACE (2004) has established an interim framework for evaluating dredged materials as shown in Figure 9.2. Chemical and biological assessments are required to test for adverse potential impacts. Engineering suitability is also required. Compactability, consolidation, and shear strength information is required. Particle size suitability has been established in Wisconsin Administrative Code, Natural Resources Chapter 347, for beach nourishment. The average silt plus clay content (#200 sieve) in the dredged material must not exceed the existing beach material by more than 15%. Wetland

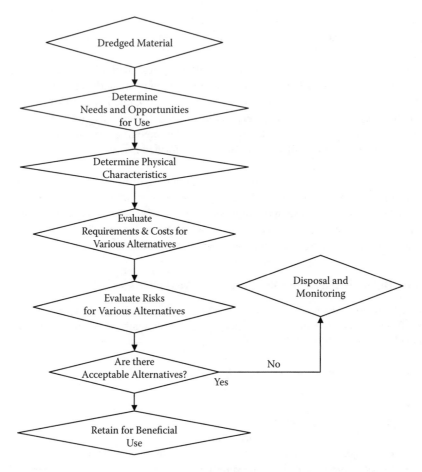

FIGURE 9.2   Evaluating the beneficial use of dredged or removed contaminated sediments (based on USACE/USEPA, 2004).

creation, upland habitat, and fisheries improvement are environmental enhancements (PIANC, 1992). Table 9.1 summarizes some of the properties required for beneficial use.

Treatment is the next step. Many methods are new and have not been tested for dredged materials. Clesceri et al. (2000) provided a treatment train for evaluation of remedial options. In general, low-temperature treatments, such as soil washing or chemical treatment, are used for low-level contamination, thermal desorption or solvent extraction for intermediate contamination, and plasma torch, rotary kiln, or fluidized sand for highly contaminated materials.

Phytoreclamation (Price and Lee, 1999) involves the use of plant-based remediation of dredged materials. Physical-chemical characteristics of the material, exposure effects on the plants and treatment efficiency should be determined. A cost-benefit assessment indicated that phytoreclamation can be a suitable treatment for restoring the sediments for beneficial reuse (Seidel et al., 2004).

TABLE 9.1

Desirable Characteristics for Dredged Materials According to Beneficial Use Option

Beneficial Use	Sediment Type				
	Rock	Gravel and Sand	Consolidated Clay	Silt/Soft Clay	Mixture
Engineering Uses					
Capping		X	X		X
Construction materials	X	X	X	X	X
Creation of berms	X	X	X		X
Fill replacement	X	X			X
Land creation	X	X	X	X	X
Land improvement	X	X	X	X	X
Shore protection	X	X	X		X
Agricultural Uses					
Aquaculture			X	X	X
Top soil				X	X
Environmental Enhancement					
Beach nourishment		X			
Fishery improvement	X	X	X	X	X
Wetland restoration			X	X	X
Wildlife habitats	X	X	X	X	X

Source: Adapted from USACE, 2007.

New beneficial uses are being evaluated. Restoration of an acid mine drainage site into a recreational park and passive remediation facility is an example (Lee et al., 2007). The dredged material was combined with waste paper fiber and processed cow manure. A constructed wetland was established with the material. Six neutralization ponds and seven acres of wetland were constructed.

The environmental aspects are mainly examined. However, physical, chemical, and biological evaluation and engineering suitability for the various options must be established. Guidelines exist for what is currently acceptable regarding the testing of the dredged material and allowable uses based on the analysis. If adverse affects are obtained, then the materials should be treated. If there is no potential for adverse impact, then beneficial use is possible. One of the main barriers is a lack of full-scale technologies. Cost information is also lacking.

General aspects for evaluation of beneficial use alternatives include benefits to humans and ecology, compatibility with the estuary or watershed, feasibility, cost, funding availability, environmental impact, regulatory requirements, public reaction, and risk. Multiple lines of evidence should be established and integrated. There is clearly a need for establishment of beneficial use alternatives for contaminated sediments. Most treatments have only been evaluated at pilot scale. Guidance and decision support tools are needed to enhance the beneficial use of dredged material.

In addition, the evaluation of the social aspects, in particular, is not standardized but is required during environmental assessment of the dredged materials. Bray (2008) described the construction of an island from dredged materials for wading birds. However, more birds arrived than predicted, and the fish population was significantly reduced. The fishermen in the area were affected and lodged many complaints, resulting in removal of the island.

9.4 SUSTAINABILITY EVALUATION

Various attempts have been made to evaluate the sustainability of remediation projects. In general, longer-term effects need to be evaluated for sustainability. DETR (1998) indicated that a project must ensure that present and future generations will be able to use the resources (sediments) and thus to continue development. The accurate prediction of the project impacts is necessary to determine sustainability. Uncertainty must be incorporated, because many consequences may be unknown or uncertain (Brooke, 1998). This, in effect, is the precautionary principle, which indicates that nothing should be done if it could cause potential damage to the resources or area. This has been incorporated increasingly into legislation (Santillo et al., 1998). Some environmental impacts include changes to the hydrodynamics and erosional characteristics at the site, impact of the work itself such as noise and disruption to the community, and loss of habitat at the site or disposal area.

Adverse effects must be avoided and must not be greater than the beneficial aspects of the project. In addition, the input of contaminants at the source must be reduced. If the source is an industrial plant in another country, this may be difficult. However, mitigation processes such as MNR cannot be sustainable if there is a continuous source of contaminants. The project itself should not lead to significant impact on the environment such as changes in water flow or increases in the suspended solids content from dredging actions. The choice of dredging equipment or the season of the dredging can have a significant environmental impact on a project. Installation of a silt curtain may prevent the release of the suspended solids into other locations, but this is only feasible in sheltered or inland areas.

Avoidance of impact on the receptor is another alternative if the effect of the origin cannot be reduced effectively. If there is a rare fish population in a lake where the project is, then that fish population could be temporarily moved to another lake. After the project is completed, the site can be altered, creating a different habitat environment. The loss of the material removed during dredging may need to be compensated for, or the habitat must be recreated by another means. Many environmental agencies thus require that the habitat must be replaced or that there be no net loss of material. Underwater berms can be created to allow spawning.

Coastal areas are particularly sensitive to the effects of climate change due to rising sea levels, which have increased by 3 mm per year in recent years. In the future, this rate could increase up to 28 to 43 cm over the next 100 years (IPCC, 2007). More construction materials will be required for berms and dikes to protect from flooding. Dredging for maintaining water levels and as compensation for increased erosion will increase. Other changes due to climate are related to salinity and temperature changes. Climate change therefore needs to be incorporated into sustainability assessments. Countries with colder climates tend to have a lower biodiversity and are thus more susceptible to significant changes. Climate can thus be a major consideration when evaluating the impact on the environment and will thus vary from country to country.

In the past, sediment managers and decision makers have had to make decisions regarding sediment management for river basins, and coastal and marine environments. Risks to the environment, health issues, and costs must be balanced, but should not be done alone. Partnering with public participation is essential and can proceed, according to Bray (2008), as follows:

- Identification of interested parties
- Submission of requests for participation in the project
- Allowing for active participation of all parties as stakeholders
- Proceeding with an environmental assessment of the project
- Allowing consultation in all parts of the project, particularly during the assessment
- Dissemination of all information on the projects
- Involvement of the end users throughout the project
- Providing open communication
- Balancing environmental and commercial goals of the project

The formation of partnering and technical advisory groups can be useful to ensure the assessment and management of the project. They can be made up of local community and environmental groups. For example, as part of the sustainable development strategy for the protection and rehabilitation of the St. Lawrence River in North America, ZIP (zone d'intervention prioritaire) committees have been established. Fourteen zones have been established along the river. As an example, the Ville Marie ZIP committee (created in 1996) covers the zone of the city of Montreal to the west of Boulevard Saint-Laurent. The Kanahwake reserve and the city of Longueuil are included on the south shore. The St. Lawrence River, the islands and the Lachine Canal, and the old Port of Montreal are included.

The Comité ZIP Jacques-Cartier was created in 1994 as the result of various people interested in the quality of the St. Lawrence River near Montreal between the Victoria Bridge and the area where the Rivière des Prairies meets the St. Lawrence downstream. The river includes the islands of Sainte-Hélène, Notre-Dame, Verte, La Batture, Haynes, and Bonfoin, as well as the islands of Boucherville and the archipel of Sainte-Thérèse. The territory is very urbanized and includes many problems related to contaminated sediments, wastewater effluents, and diffuse pollution from the Port of Montreal, Sector 103.

The Comité ZIP Jacques Cartier is a nonprofit organization that groups the different interests for the purpose of resolving problems that involve the St. Lawrence River. In 1993, oil-contaminated sediment started to float to the surface due to disturbance by heavy ships. Environment Canada emergency crews had to intervene seven times in three months to remove floating oil. Since 1994, the Comité ZIP played a role with the community in implementing the governmental program Saint-Laurent Vision 2000 (SLV 2000, http://www.slv2000.qc.ec.gc.ca/index_a.htm). The Comité ZIP Jacques Cartier consists of representatives from industry, citizens, environmental, socioenvironmental, and municipal groups. They were instrumental in the recent remediation effort at Pier 103 of the Port of Montreal. They coordinated the consulting group as well as the monitoring committee. The community was involved for more than 10 years in the project. Various levels of government, the provincial (MDDEP), federal (Environment Canada), and the City of Montreal collaborated, in addition to all workers in the field. The field work was transparent and open (Figure 9.3). The results of the project were previously discussed in Chapter 7. The three companies and the Port of Montreal assumed the whole cost of the $10 million project. This was necessary because there are no other provincial or federal programs to assist in aquatic site remediation. More than 98.5% of the petroleum hydrocarbons, copper, and selenium and 99% of the PAHs were removed (www.grouperestauration103.com/articles). After biological treatment of the oil-contaminated sediments was employed, residual material was either used as fill at one of the industrial sites (Imperial) or sent to a

FIGURE 9.3 Photographs of sediment removal and silt curtain during restoration of Sector 103 in Montreal, Canada.

soil treatment and landfill site in the Mauricie region of Quebec, thus minimizing transport. Partnership and community were key elements of the project, as is an example of the integration shown in Figure 9.4.

As indicated by Bray (2008), a major problem is to find a balance between economics and the environment effects and benefits of a project. It is also easy to show costs but not as easy to indicate environmental effects. Some environmental aspects can be indicated economically, such as loss of property values, loss of fishing species, and increased health care costs. However, the loss of a species with no commercial value is very difficult to assess. To move toward sustainable development and optimal use of the resource economics is a necessary element in the evaluation. In addition, the restoration of the initial state of the ecosystem must be achieved, as shown in Figure 9.5, but may not be reached by technological mitigation.

9.5   CASE STUDY OF LACHINE CANAL

In Canada, the Lachine Canal, Montreal, Canada, was the center of industrial activity in the early 1800s. It was established to circumvent the Lachine Rapids, which was required to connect the Great Lakes with the St. Lawrence River. It thus served as a link between the Atlantic and the middle of the continent. The Canal also had an influence on the manufacturing industry and the urbanization of the island of Montreal. It is 14.5 km in length and stretches from the Old Port of Montreal to Lake Saint-Louis. It operated for 175 years, when it was closed in 1959 due to the

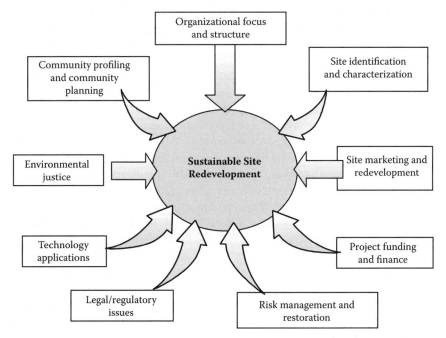

FIGURE 9.4   Elements to be incorporated into a sustainable site redevelopment (adapted from Yong et al., 2006).

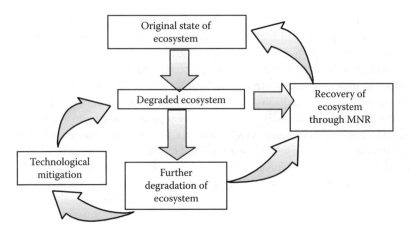

FIGURE 9.5 Ultimate recovery of the original ecosystem state through various approaches.

opening of the St. Lawrence Seaway. Because it is located near the Port of Montreal, many industries were established along the canal. Unfortunately many of these have contributed to the contamination of the canal. From 1978, Parks Canada forbade the use of the canal for recreational purposes until 1992, when pleasure boating was reintroduced. Millions of cyclists and pedestrians also visit the banks of the Canal (Figure 9.6).

Environmental impact studies were performed and indicated that the contamination reduced the value of the site. Municipal discharges also led to elevated bacterial levels in the canal. As part of the St. Lawrence Action Plan, the decontamination of the canal was launched. The contamination was mainly in the top 1 m of the sediment, and the volume of contaminated sediments was estimated at 215,000 m^3.

Environment Canada requested the creation of a Commission for the environmental evaluation of the decontamination of the Lachine Canal. The examination had to be a federal–provincial study, because a portion of the canal was under provincial jurisdiction. In May 1991, the Commission mandated that different options be evaluated for the decontamination and that the most viable be presented. The impact study of the decontamination was completed in 1993. This was followed by the reopening of the Canal for pleasure craft in 2002.

A series of analyses were performed on the sediment in the Canal. SNC-Lavalin performed a study in 1992 in water of Lake St. Louis, the Bay of Lachine, and in the canal, as well as the sediments in the canal and the Bay. The CCME criteria for Cd, Cr, Cu, and Pb in water were slightly exceeded for the protection of aquatic life. The concentration of the metals, Cd, Cu, and Pb, exceeded the St. Lawrence and Centre Saint-Laurent criteria (threshold effect level, TEL) in the sediments of the Bay of Lachine. In the Canal, elevated levels of Hg, Pb, Zn, and high-molecular-weight PAHs and PCBs were found. The study found that an intervention was required, because the SEN levels of the Centre Saint Laurent were exceeded.

GPR International (1992) determined the volume of sediments by a geophysical approach. Seismic refraction, georadar, bathymetry, and sonar were used to

FIGURE 9.6   Photographs of the Lachine Canal, Montreal, Canada.

qualitatively and quantitatively characterize the sediment. The volume of sediments was estimated at 121,000 m^3 with a margin of error of approximately 40,000 m^3. The layer of sediments was an average of 26 cm in thickness. Downstream, a zone in the canal was identified as a priority zone for intervention.

To prepare for the new navigation in the canal, Parks Canada (Parcs Canada, 1998) took various samples with a grab sampler. Of the 24 samples taken, 18 were analyzed for metals, C10–50 hydrocarbons, PAH, and PCB (Aroclors and total); most exceeded the interim values for major effects of the Centre Saint Laurent (TEL) level for metals Cd (3 ppm), Cu (86 ppm), Pb (170 mg/kg), and Zn (540 mg/kg), or PAH. Most of these were probably removed by excavation.

In the past, various studies have been performed to evaluate technologies for the treatment of the site (SNC Lavalin Environnement, 1992). Physicochemical, incineration, solidification/stabilization, and biological categories were evaluated. According to technical and economical criteria, two approaches were feasible: solidification/stabilization and washing. Nine promoters were selected for comparison. Laboratory tests were then performed for seven of the technologies (Environment Canada, 1992). Six were based on physical or chemical fixation, while one was based on biodegradation–biofixation.

An evaluation by Tecsult in 1993 (Tescult et Roche, 1993) was performed to determine the risk to human health by contact with water, sediment by oral intake or dermal contact, and for the consumption of fish. Risk due to fish intake was seen, but more studies were needed. It was, therefore, decided to forbid the consumption of fish from the Lachine Canal. Other risks were minimal. Another study by Tecsult in 1993 evaluated the suspended solids in the water column. By using a probable level of contamination, the levels of contamination in the sediment would not pass the SEN level within the next 100 years. However a more pessimistic scenario indicated that zinc and mercury might pass the levels in the next 70 and 20 years, respectively, at the current level of sedimentation.

Appropriate professional and technical consultation and monitoring measures are required to ensure that the heritage of the site is protected in addition to safety aspects. Dredging operations might lead to resuspension of the contaminated sediments which could be carried downstream.

In 1994, dredging and the status quo were the main options to be examined by the Commission for the protection of benthic organisms. A joint federal–provincial commission studied the decontamination of the Lachine Canal in 1996. According to Parks Canada and the Old Port of Montreal, encapsulation was seen as the best option. However, the Commission recommended nonintervention and considered that encapsulation was not acceptable due to the technical unknowns and economic aspects. The risks of resuspension were to be evaluated. Finally, Parks Canada in 1996 presented a review of the reports of the restoration.

The through traffic itself should not cause contamination, unless sanitary effluents are discharged into the canal. Appropriate measures for fuel should also be followed. Tecsult (Parcs Canada, 1998) performed a study on the risk of resuspension of sediments due to boat navigation and lock operation and concluded that the effects should be minimal. Various scenarios based on the type of boat and speed were evaluated. In 2002, Parks Canada initiated a monitoring program during the navigation season to evaluate if the sediments were stirred up in the Lachine Canal. The negative impacts will need to be mitigated by known techniques.

In 1999, Environment Unlimited (Environnement Illimité inc., 1999) performed another study of the effect of pleasure craft on the suspension of the sediments. Speeds of 5 to 10 km/hour were evaluated by tests at three different sites. No resuspension was observed. However, at 15 km/hour, resuspension occurred, but did not affect more than 1 m of the water column. Beginning in 2002, the quality of the suspended solids was monitored. Rarely were elevated levels measured. Sediment traps were used for determining the content of sediments for metals and PAHs. Some exceedances of the TEL levels for various metals (Cu, Pb, Zn, and PAHs) have been determined.

In 2006, Dessau Soprin performed an analysis to evaluate the effect of the opening of the canal on surface sediments. It was found that the surface sediments were contaminated with PAH and metals, in particular. PCBs were not problematic. Dessau (2006) indicated that it was not possible to determine the reason for the increased concentrations since 1992. Some potential causes are contaminant sources along the Canal, the remobilization due to the pleasure craft, or differences in methods of sampling or analysis. The suspended solids levels are rarely more than 4 mg/L and the concentration of the metals in these solids is lower than the sediments. Therefore, it is unlikely that this is the cause of the enhanced concentration of contaminants in the sediments. Improvements in laboratory analyses or the slight variations in the sampling locations might account for the differences. These hypotheses have to be evaluated. Some work has been performed at some sites. Another study is to be performed in 2009–2010 with sampling to be performed to determine the risk assessment of leaving the contaminants in place. This project is a key example of difficulties caused by a lack of economic alternatives.

9.6 BARRIERS TO TECHNOLOGY DEVELOPMENT AND IMPLEMENTATION

Disposal in landfills is often the least expensive alternative. If disposal is not as secure or not accessible, then partial decontamination may be feasible. Economics is thus a major consideration. Regulatory and public acceptance are other issues. Integration of treatment technologies in an overall sediment management scheme is often not practiced. However, due to the varied nature of the sediments, this is highly desirable. Contaminated sediments usually contain a variety of heavy metals, petroleum, and chlorinated hydrocarbons. Many treatments are still at the pilot, or demonstration stage. In situ methods are effective and low cost. Natural recovery (NR) is most appropriate for low-level contamination. Capping may be used if NR is not effective.

Source control is an absolute requirement. Long-term monitoring will assist in determining the effectiveness of source control and the mitigation strategy, particularly degradation and contaminant release. All must be done on a site-by-site basis, with a strong scientific background. Public perception will be improved through citizen and community forums and education of the public of in situ means. Recently in a survey in Norway, the public indicated that their preference was capping (Breedveld, 2008).

Dredging procedures must be more precise to limit removal of uncontaminated sediments. Contained aquatic disposal near the site can be acceptable due to reduced transportation. However, research is needed to determine the capability for enhanced habitat improvement, and evaluate the degradation of contaminants within the cap, and identify release of the contaminants. Regarding ex situ technologies, bench and pilot tests are needed to evaluate the effectiveness of the technologies such as bioremediation. Cost data need to be improved, particularly for large-scale projects where there is the economy of scale.

Marketable products should be developed and produced to enhance process economics. Recycling of dredged products can be mandated in public projects. The public also needs to be educated concerning the benefits of sediment beneficial reuse.

9.7 CURRENT NEEDS AND FUTURE DIRECTIONS

Standardization of sampling and testing techniques is needed. For many of the tests, there is significant variability and a lack of analytical methods. Therefore, development of new methods is needed. Inexpensive screening methods for contaminants are also needed, because many are very costly. In addition, methods for analysis of the chronic/sublethal effects on organisms must be developed.

X-ray absorption spectrometry is useful for determining the distribution and bonding of metals. However, due to the sophistication and cost of the instrument, use of this technique will most likely be for research only. Further studies on environmental samples are required to determine the kinetics of release and transport of chemicals.

Selective sequential extraction (SSE) techniques are used to examine contaminant release from the surface to the pore water. It is difficult to convert this information to bioavailability, although some attempts have been made. This information is essential for evaluating remedial alternatives and effectiveness of the remediation (NRC, 2003).

For bioaccumulation to occur, contaminants must first be released from the contaminants and then absorbed by the organism. Diffusion, dispersion, and advection are the main transport mechanisms of contaminants to the organism. Some organisms are in direct contact with the sediment, whereas others absorb dissolved chemicals or eat organisms that have already absorbed contaminants. The main techniques that are available are acid volatile sulfide (AVS) and biota–soil/sediment accumulation factor (BSAF). AVS is related to metals, while BSAF is mainly related to organic compounds. Both techniques are simplistic and subject to many limitations. Data is often site and species specific. Currently no specific guidance or consensus exists regarding the use of bioavailability process assessment.

Recent indications of various locations such as the St. Lawrence River ecosystem have shown that the controls and remediation of the area by industry and environmental organizations have led to decreases of many substances such as PCBs and heavy metals, including mercury (Environment Canada and MDDEP, 2008). However, much remains to be determined regarding tributyltin (TBT) (marine paint), dioxins and furans (various industrial discharges), and new emerging substances including polybrominated diphenyl ethers (PBDEs) (flame retardants in household products) and perfluorooctanes (PFOs) and perfluorinated alkyls (PFAs) (water and oil repellents in textiles and packaging). Antibiotics from agricultural fields are other contaminants of concern. These substances pose new challenges and are of increasing priority due to the potential harmful effects on benthic organisms. They are also difficult to source control and are widespread in nature.

Active caps are simple and effective and an alternative to dredging. There is the potential for combination with treatment and habitat restoration. Research is needed to determine physical, chemical, and biological processes controlling fate and transport. Gas ebullition affects the transport of non-aqueous-phase liquids (NAPLs). Long-term aspects of stability, ice scour, erosion processes, and the diversity of the benthic community will need to be studied.

For capping projects, there is a lack of information available outside the industry regarding the capping materials for in situ capping projects. New materials are being developed and must be suitable for the environment at a reasonable cost. More data is needed from existing projects. The impact of extreme events such as storms and floods on the cap is needed to predict their behavior in the future and design better caps.

For dredging projects, more information is needed for new dredging techniques to avoid damage in ecologically sensitive areas. It can destroy benthic habitats and can lead to resuspension of sediments in neighboring areas. Subsequent disposal has technical and political problems. Other impacts such as fuel and energy consumption need to be reduced. The impact will vary according to climate and location as previously mentioned.

Natural recovery can reduce the area of contamination for dredging or treatment. Little is known about the in situ processes such as advection and biodegradation. Also, how can toxicity and bioavailability, as well as the time for remediation, be reduced? Contaminated sites have complex mixtures of pollutants. There is the potential accumulation of intermediates. However, few studies have been performed using real sediments. In addition, new mixtures such as pharmaceuticals, hormones, and endocrine disruptors are becoming high priority. Therefore, how can characterization and detection of biotransformation products and pathways be determined? Other challenges are how to extrapolate laboratory to in situ conditions, how to predict bioremediation potential, and how to optimize microorganisms that will be involved. Hydrophobic PAHs, PCBs, and dichlorodiphenyltrichloroethane (DDT) are difficult to treat. Solubilization agents such as biosurfactants could enhance biodegradation rates but must not interfere with microbial processes.

Monitoring of microbial processes (nitrate, sulfate, Fe(III), or Mn(IV), consumption, fatty acids, and methane, is important. Hydrogen production, by-product formation, nutrient consumption, and dynamics of microbial community are all parameters that should be monitored to understand the in situ biological processes more thoroughly. Laboratory studies are used as key lines of evidence for MNR, but must be used with biogeochemical evidence. Site modeling can also have great relevance, and the limitations must be understood and minimized.

Monitoring strategy development is needed, particularly for MNR but also for other sediment management strategies, because monitoring has been inadequate at most sites. The data should be openly available for evaluation of the effectiveness of the remediation. Monitoring techniques to evaluate contaminant release and biota monitoring as indicators of accumulation in the food web are required.

Overall, research into the improvement and development of new remediation techniques, site characterization, and monitoring of sediment contamination is required and should be supported. Remedies must be cost-effective. Working with public interest groups, regulatory agencies, and other consultants for field trials will assist in the development of acceptable and viable technologies. Demonstration projects must be funded to enable the development of new technologies and reduction of their risks. Evaluation of technologies at the same site would be particularly beneficial to enable comparison under similar conditions.

In situ management techniques offer many cost advantages compared to excavation and relocation of the materials. Monitoring must play an integral part to ensure long-term effectiveness. Physical, chemical, and biological processes need to be understood for MNR and other in situ approaches such as capping and chemical treatment. Biodegradation is not well understood in subaqueous and marine environments.

Regarding ex situ technologies, dredging is utilized for removal of contaminated sediments. More precise methods are being developed. Containment facilities such as confined disposal facilities (CDFs) are most appropriate for interim storage and can enhance sediment separation for potential sediment reuse. Long-term monitoring is needed to ensure that contaminant releases are controlled. Contained aquatic disposal (CAD) is used particularly for shallow sites but is not frequent. Contaminated sediments can be lost during placement of the sediments, and improved techniques are needed to estimate long-term effectiveness.

A variety of approaches have been utilized for ex situ treatment of contaminated sediments. Many have been lab or pilot tested. Marine applications, in particular, need larger-scale testing. Chemical, thermal, and immobilization techniques have been utilized but are expensive and complicated. Biological ex situ treatment has the potential to be cost-effective, but many technical issues remain, particularly concerning system design. However, factors can be controlled much more easily than for in situ processes.

A comparison of the feasibility, effectiveness, practicality, and cost of various technologies was performed by the NRC (1999). Effectiveness is based on isolation or removal efficiency; feasibility is based on the technological development (such as lab scale or commercial). Practicality is based on public acceptance, and cost is based on the cost of the method only (associated costs were not included). The results are shown in Figure 9.7 and indicate that trade-offs are required because there is no clear winner. In addition, comparison must be made on a site-specific basis.

Various substances have been discharged into enclosed water areas such as lakes, ponds, and the enclosed sea areas from the shores and rivers by natural and human activities. Suspended solids (SS) consisting of inorganic and organic substances have been discharged into water areas. It was reported by Hoshika et al. (1996) that the amount of SS discharged was over 200,000 tons during the summer in Osaka Bay in Japan. In addition, the amount of organic particles in the SS was approximately 150,000 tons (Hoshika et al., 1996). These SS which exist in the water cause

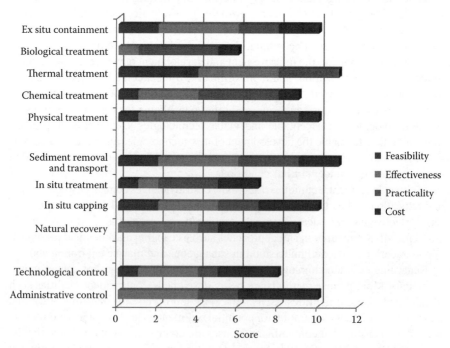

FIGURE 9.7 Comparison of various technologies for sediment management (data from NRC, 1999).

a decrease in transparency and a reduction of dissolved oxygen. The reduction of dissolved oxygen in the bottom water and the sediment causes the anaerobic condition and results in the elution of reduction products such as heavy metals and nutrients from the sediment. de Jong (2006) also noted that, in the Netherlands, suspended solids from dredging and other activities is a substantial problem because the solids block light transmission and subsequent photosynthetic processes. Approaches such as previously described for control of suspended solids will require further development to protect sediment quality.

In addition, nutrient leaching from sediments can lead to eutrophication in the water. It is one of the biggest environmental problems in enclosed water areas. It was noted by the World Health Organization (1999) that, in the Asia Pacific Region, 54% of lakes are eutrophic. The proportions for Europe, Africa, North America, and South America are 53%, 28%, 48%, and 41%, respectively. Thus, eutrophication is a common and serious problem around the world. Algal blooms are also important problems for the environment in enclosed water areas (Codd et al., 2005). Phosphorus (P) plays a key role in the eutrophication state of surface water. Sediments are an important source that provide relatively high loads of P during the recovery period of the water body after the external loads have been reduced. Adsorption–desorption processes are involved in the equilibrium between the sediments and the overlaying water. These processes are not well understood. This state is determined by the equilibrium conditions between the water column and sediments interface. The sediments can act as a source of P, and they must be considered during development of a management scheme for the surface water.

9.8   CONCLUDING REMARKS

The health of the aquatic geoenvironment is vital to the productivity of aquatic resources. Surface water environments (fresh water and marine) are significant resource bases and are essential components for humans. Man-made events result in the discharge of wastes from ocean vessels and from land-based industries and activities. Although discussion of global warming has not been extensive in this book, it is important to point out that aquatic geoenvironment will be affected by this.

Contaminants accumulate in the sediments, while others are found in the food chain through bioaccumulation, thus posing health threats to humans and biota. Eutrophication and concentration of toxic and hazardous substances are considerable problems. Hazardous substances and nutrients have been discharged into the aquatic environment for many decades. These have to be collected and removed. A balance in the amount of nutrients removed is needed. On the one hand, sufficient removal of excess nutrients is needed to avoid eutrophication, and on the other hand, sufficient nutrients must be available for the aquatic animals that rely on these nutrients for their food supply. A sustainable aquatic geoenvironment requires an integrated approach. Research and development into how this can be achieved would contribute substantially. Without proper control and management of pollution sources, mitigation of sediment contamination cannot be achieved. This includes the management of suspended solids in the water column. Hot spots and low-level contaminated sediments must be managed in a low-cost and least-disruptive manner to achieve the ultimate goal of full restoration of the aquatic geoenvironment. Monitoring has not been extensively employed to evaluate and develop a better understanding of the effect of remedial technologies, but is the only means of evaluating success in achieving the ultimate remedial goals and sustainability of the remediation.

A long-term vision is needed. Otherwise, natural resources will continue to be depleted, waste will not be minimized, landfill will continue to be filled with contaminated sediments, and benthic and wetland habitats will be lost. Innovative decontamination technologies with beneficial use need to be developed and applied through regulatory encouragement and financial mechanisms.

REFERENCES

Bray, R.N. (Ed.). 2008. *Environmental Aspects of Dredging*. Taylor and Francis, Leiden, The Netherlands.

Breedveld, G. 2008. Challenges in remediation of contaminated sediments in Norway. *5th SedNet Conference,* May 27–29, 2008, Oslo, Norway.

Brooke, J. 2007. Strategic coastal defence planning: The role of the planning system. *Water Env. J.* 14(2): 140–142.

Clesceri, N.L., Stern, E.A., Feng, H., and Jones, K.W. 2000. Decontaminating and processing dredged material for beneficial use. BNL-67484, Brookhaven National Lab, Upton, NY.

Codd, G.A., Morrison, L.F., and Metcalf, J.S. 2005. Cyanobacterial toxins: Risk management for health protection. *Toxicol. Appl. Pharm.* 203(3): 264–272.

de Jong, V.N. 2006. Functions of mud in estuarine and coastal ecosystems. *4th SedNet Conference,* 23–24 November 2006, Venice, Italy, Nov.

Dessau Soprin Inc. 2006. *Caractérisation environnementale des sédiments*, Canal de Lachine, Québec, Rapport final for the Travaux public et services gouvernementaux Canada (TPSGC) et Parcs Canada, Mars 2006, Poo4812-100-SE-0100-00.

DETR. 1998. Sustainable local communities for the 21st century: Why and how to prepare an effective Local Agenda 21 strategy. Department of the Environment, Transport and the Regions, UK.

Environment Canada. 1999. Canadian Environmental Protection Act, Part 7, Division 3, Schedules 5 and 6 and the Disposal at Sea Regulations.

Environnement CAnada, avril 1992. Validation de procédés de traitement des sédiments contaminés du canal de Lachine, rapport final. Division des technologies de restauration, Centre Saint-Laurent.

Environment Canada and MDDEP. 2008. *State of the St. Lawrence River, Toxic Contamination in Sediments, Lake St. Louis: Where Two Rivers Meet*, 11 pp.

Environnement Illimité inc. 1999. Canal Lachine-Étude des effets de la navigation de plaisance sur la remise en suspension des sédiments. Rapport final, 39 pp.

Géophysique GPR International Inc. 1992. Canal de Lachine—Mesure du volume de sédiments par l'approche géophysique—Modélisation de type interactif, dynamique. Rapport final en 7 volumes, N réf M-92272 p.

Hoshika, A., Tanimoto, T., and Mishima, Y. 1996. Characteristic of particulate matter in Osaka Bay, the Seto Inland Sea. *Reports of Chugoku National Industrial Research Institute* 47, 15–26.

IPCC. 2007. Intergovernmental Panel on Climate Change. Climate 2007, WMO, UNEP, IPCC Secretariat, c/o WMO Geneva, Switzerland.

Jim Osborne Consulting. 2006. The contaminated dredged materials management decisions workshop, Montreal, November 27–30, 2006. Final report. Prepared for Environment Canada.

Lee, C.R., Brandon, D.L., and Price, R.A. 2007. Manufactured soil field demonstration for constructing wetlands to treat acid mine drainage on abandoned minelands. DOER Technical Notes ERDC TN-DOER-D9. U.S. Army Engineer Research and Development Center, Vicksburg, MS.

Ludwig, D., Truchon, S., and Tammi, C. 2006. A Framework for River Cleanup Decision Making, foundation principles for managing the Neponset River Watershed. In E.J. Calabrese, P.T. Kostecki, and J. Dragun (Eds.), *Contaminated Soils, Sediments and Water, Successes and Challenges,* Vol. 10, Springer US, New York, pp. 359–366.

NRC. 1999. *National Symposium on Contaminated Sediments, Coupling Risk Reduction with Sustainable Management and Reuse.* National Research Council. National Academy Press, Washington, DC.

NRC. 2003. *Bioavailability of Contaminants in Soils and Sediments: Processes, Tools and Applications.* National Research Council, National Academies Press, Washington, DC.

Parcs Canada. 1998. Examen prélable : Projet de réouverture du Canal de Lachine à la navigation de transmit. Rapport final, 115 pp.

PIANC. 1992. Beneficial uses of dredged material. A practical guide. PIANC Report of PIANC Working Group 19, Permanent International Association of Navigational Congress.

Price, R.A. and Lee, C.R. 1999. Evaluation of dredged material for phytoreclamation suitability. DOER Technical Notes Collection TN-DOER-C3, Vicksburg, MS. U.S. Army Engineer Research and Development Center.

Santillo, D. 1998. The precautionary principle: Protecting against failures of scientific method and risk assessment. *Marine Poll. Bull.* 36(12): 939–950.

Seidel, H., Loser, C., Zehnsdorf, A., Hoffmann, P., and Shmerold, R. 2004. Bioremediation process for sediments contaminated by heavy metals feasibility study on a pilot scale. *Environ. Sci. Technol.* 38(5): 1582–1588.

SNC Lavalin Environnement. 1992. Canal de Lachine—Caractérisation de l'eau et des sédiments. Rapport final—Volume 1, N réf 57391, 124 p.

Tecsult, mai 1993. Suivi des matiéres en suspension à l'été 1992. Document de support, évaluation environnementale, project de décontamination du Canal de Lachine, pour le compte de Parcs Canada.

Tecsult et Roche. 1993. *Évaluation environnementale, Projet de décontamination du canal de Lachine,* for Parks Canada and the Old Port of Montréal Corporation, September 1993, 2 vol.

Thibodeaux, L.J., Duckworth, K., Reible, D.D., and Valsaray, K.T. 1999. Effective of environmental dredging, finding from case studies. *Proc. 20th Annual Meeting of Environmental Toxicology and Chemistry (SETAV),* Nov. 14–18, 1999, Philadelphia. SETA, Pensacola, FL.

USACE. 2007. Dredging Operation and Environmental Research Program. Summary of Available Guidance and Best Practices for Determining Suitability of Dredged Material for Beneficial Uses. ERDC/EL TR-07-27, U.S. Army Corps of Engineers, Washington, DC.

USACE/USEPA. 2004. Evaluating environmental effects of dredged material management alternatives—a technical framework. EPA 842-B-92-008, U.S. Army Corps of Engineers and U.S. Environmental Protection Agency, Washington, DC.

World Health Organization (WHO). 1999. Toxic *Cyanobacteria in Water: A Guide to Their Public Health Consequences, Monitoring and Management.* Chorus, I. and Bartram, J. (Eds.), Spon Press, London, 416 pp.

Yong, R.N., Mulligan, C.N., and Fukue, M. 2006. *Geoenvironmental Sustainability.* CRC Press, Boca Raton.

Appendix A: Sediment Quality Guidelines from Environment Canada and MDDEP, 2008

TABLE A.1

Criteria for the Sediment of Freshwater Sediment Quality in Quebec

Group	Substance	Concentrations (mg/kg)[a,b]				
		REL	TEL	OEL	PEL	FEL
Metals and Metalloids	Arsenic	4.1	5.9	7.9	17	23
	Cadmium	0.33	0.60	1.7	3.5	12
	Chromium	25	37	57	90	120
	Copper	22	36	63	200	700
	Lead	25	35	52	91	150
	Mercury*	0.094	0.17	0.25	0.49	0.87
	Nickel	ND	ND	47	ND	ND
	Zinc	80	120	170	310	770
Organic Compounds	Total polychlorinated biphenyls (PCBs)*	0.025	0.034	0.079	0.28	0.78
	Nonylphenol and its ethoxylates[c]	ND	1.4	ND	ND	ND
	PCDD/PCDF (ng tox eq/kg)*[d]	0.27	0.85	10	22	36
Polycyclic Aromatic Hydrocarbons	Acenaphthene[e]	0.0037	0.0067	0.021	0.089	0.94
	Acenaphthylene[e]	0.0033	0.0059	0.030	0.13	0.34
	Anthracene[e]	0.016	0.047	0.11	0.24	1.1
	Benzo[a]anthracene	0.014	0.032	0.12	0.39	0.76
	Benzo[a]pyrene	0.011	0.032	0.15	0.78	3.2
	Chrysene	0.026	0.057	0.24	0.86	1.6
	Dibenzo[a,h]anthracene[e]	0.0033	0.0062	0.043	0.14	0.20
	Fluoranthene	0.047	0.11	0.45	2.4	4.9
	Fluorene[e]	0.010	0.021	0.061	0.14	1.2
	2-Methylnaphtalene[e]	0.016	0.020	0.063	0.20	0.38
	Naphtalene[e]	0.017	0.035	0.12	0.39	1.2

Phenanthrene	0.025	0.042	0.13	0.52	1.1
Pyrene	0.029	0.053	0.23	0.88	1.5
Organochlorine Pesticides					
Chlordane	0.0015	0.0045	0.0067	0.0089	0.015
DDD*f	0.00035	0.0035	0.0085	0.0085	0.015
DDE*g	0.00025	0.0014	0.0026	0.0068	0.019
DDT*h	0.00033	0.0012	0.0038	0.0048	0.010
Dieldrin	0.00044	0.0029	0.0039	0.0067	0.017
Endrin	0.00063	0.0027	0.036	0.062	0.33
Heptachlor epoxide	0.00026	0.00060	0.0027	0.0027	0.0040
Lindane	0.00022	0.00094	0.0014	0.0014	0.011
Toxaphene*i	ND	0.00010	ND	ND	ND

Note: REL: rare effect level; TEL: threshold effect level; OEL: occasional effect level; PEL: probable effect level; FEL: frequent effect level; ND: not determined.

* For these persistent, bioaccumulative, and toxic substances (SLV 2000 1999), bioaccumulation effects may be observed in aquatic, avian, and terrestrial consumers at various trophic levels. These effects are not taken into consideration in the quality area presented here.

a The values have been rounded to two significant digits. The TEL and PEL columns contain CCME values and the REL, OEL, and FEL columns the additional reference values.

b All the values are expressed in milligrams per kilogram (mg/kg) of dry sediment, except for the PCDD/PCDF values, which are expressed as nanograms per kilogram (ng tox eq/kg).

c Value determined by CCME (2002a) using the equilibrium partitioning method and assuming a total organic carbon (TOC) level of 1% the calculation is based on toxicity equivalency factors.

d PCDD/PCDF: polychlorinated dibenzo-p-dioxin/polychlorinated dibenzofurans, values are expressed in toxicity equivalency units (1). In accordance with the CCME protocol, the initial values obtained during the calculation of quality criteria were divided by a factor of 10.

e The values calculated for marine sediments were adapted by default.

f DDD: 2, 2-bis(p-chlorophenyl)-1,1-dichloroethane or dichlorodiphenyldichloroethane. This criterion applies to the sum of the p,p' and o,p' isomers.

g DDE: 1, 1-dichloro-2, 2, bis(p-chlorophenyl)ethylene or dichlorodiphenyldichloroethylene. This criterion applies to the sum of the p,p' and o,p' isomers.

h DDT: 2,2-bis(chlorophenyl)-1,1,1-trichloroethane or dichlorodiphenyltrichloroethane. This criterion applies to the sum of the p,p' and o,p' isomers.

i New York State Department of Environmental Conservation (1994) value adopted by CCME (2002b). The value was derived by using the equilibrium partitioning method and assuming a total organic carbon (TOC) level of 1%.

TABLE A.2

Criteria for the Sediment of Marine Sediment Quality in Quebec

Group	Substance	Concentrations (mg/kg)[a,b]				
		REL	TEL	OEL	PEL	FEL
Metals and Metalloids	Arsenic	4.3	57.2	19	42	150
	Cadmium	0.32	0.67	2.1	4.2	7.2
	Chromium	30	52	96	160	290
	Copper	11	19	42	110	230
	Lead	18	30	54	110	180
	Mercury*	0.051	0.13	0.29	0.70	1.4
	Nickel	ND	ND	ND	ND	ND
	Zinc	70	120	180	270	430
Organic Compounds	Total polychlorinated biphenyls (PCBs)*	0.012	0.022	0.059	0.19	0.49
	Nonylphenol and its ethoxylates[c]	ND	—	ND	ND	ND
	PCDD/PCDF (ng tox eq/kg)[d]	0.27	0.85	10	22	36
Polycyclic Aromatic Hydrocarbons	Acenaphtene	0.0037	0.0067	0.021	0.089	0.94
	Acenaphthylene	0.0033	0.0059	0.031	0.13	0.34
	Anthracene	0.016	0.047	0.11	0.24	1.1
	Benzo[a]anthracene	0.027	0.075	0.28	0.69	1.9
	Benzo[a]pyrene	0.034	0.089	0.23	0.76	1.7
	Chrysene	0.037	0.11	0.30	0.85	2.2
	Dibenzo[a,h]anthracene	0.0033	0.0062	0.043	0.14	0.20
	Fluoranthene	0.027	0.11	0.50	1.5	4.2
	Fluorene	0.010	0.021	0.061	0.14	1.2
	2-Methylnaphthalene	0.016	0.020	0.063	0.20	0.38
	Naphthalene	0.017	0.035	0.12	0.39	1.2
	Phenanthrene	0.023	0.087	0.25	0.54	2.1

Pyrene	0.041	0.15	0.42	1.4	3.8
Chlordane	0.00092	0.0023	0.0033	0.0048	0.016
Organochlorine Pesticides					
DDD[e]	0.00063	0.0012	0.0040	0.0078	0.028
DDE[f]	0.00079	0.0021	0.074	0.37	0.56
DDT[g]	0.00033	0.0012	0.0038	0.0048	0.010
Dieldrin*	0.00038	0.00071	0.0020	0.0043	0.0060
Endrin[h]	0.00063	0.0027	0.036	0.062	0.33
Heptachlor epoxide	0.00026	0.00060	0.0027	0.0027	0.0040
Lindane	0.00022	0.00032	0.00051	0.00099	0.0119
Toxaphene[i]	ND	0.00010	ND	ND	ND

Note: REL: rare effect level; TEL: threshold effect level; OEL: occasional effect level; PEL: probable effect level; FEL: frequent effect level; ND: not determined.

* For these persistent, bioaccumulative, and toxic substances (SLV 2000 1999), bioaccumulation effects may be observed in aquatic, avian, and terrestrial consumers at various trophic levels. These effects are not taken into consideration in the quality area presented here.

ᵃ The values have been rounded to two significant digits. The TEL and PEL columns contain CCME values and the REL, OEL, and FEL columns contain the additional reference values.

ᵇ All the values are expressed in milligrams per kilogram (mg/kg) of dry sediment, except for the PCDD/PCDF values, which are expressed as nanograms per kilogram (ng tox eq/kg).

ᶜ Value determined by CCME (2002a) using the equilibrium partitioning method and assuming a total organic carbon (TOC) level of 1% the calculation is based on toxicity polychlorinated dibenzofurans, values are expressed in toxicity equivalency units (1). In accordance with the CCME protocol, the initial values obtained during the calculation of quality criteria were divided be a factor of 10.

ᵈ PCDD/PCDF: polychlorinated dibenzo-*p*-dioxin/polychlorinated dibenzofurans, values are expressed in toxicity equivalency units (1). In accordance with the CCME protocol, the initial values obtained during the calculation of quality criteria were divided be a factor of 10.

ᵉ DDD: 2, 2-bis(*p*-chlorophenyl)-1,1-dichloroethane or dichlorodiphenyldichloroethane. This criterion applies to the sum of the *p,p′* and *o,p′* isomers.

ᶠ DDE: 1, 1-dichloro-2, 2, bis(*p*-chlorophenyl)ethylene or dichlorodiphenyldichloroethylene. This criterion applies to the sum of the *p,p′* and *o,p′* isomers.

ᵍ DDT: 2,2-bis(chlorophenyl)-1,1,1-trichloroethane or dichlorodiphenyltrichloroethane. This criterion applies to the sum of the *p,p′* and *o,p′* isomers.

ʰ The values calculated for freshwater sediments were adopted by default.

ⁱ New York State Department of Environmental Conservation (1994) value adopted by CCME (2002b). The value was derived by using the equilibrium partitioning method and assuming a total organic carbon (TOC) level of 1%.

REFERENCES

CCME. 2002a. Canadian Sediment Quality Guidelines for the Protection of Aquatic Life: Nonylphenol and its ethoxylates, in Canadian Environmental Quality Guidelines, 1999, Winnipeg, Manitoba.

CCME. 2002b. Canadian Sediment Quality Guidelines for the Protection of Aquatic Life: Toxaphene, in Canadian Environmental Quality Guidelines, 1999, Winnipeg, Manitoba.

New York State Department of Environmental Conservation. 1994. Technical guidance for screening contaminated sediments. Prepared by the Division of Fish and Wildlife and the Division of Marine Resources, Nov. 22, 1993, New York.

Appendix B: London Convention and Protocol: Convention on the Prevention of Marine Pollution by Dumping of Wastes and Other Matter, 1972

http://www.imo.org/Conventions/contents.asp?topic_id=258&doc_id=681
Adoption: 13 November 1972
Entry into force: 30 August 1975

INTRODUCTION

The Inter-Governmental Conference on the Convention on the Dumping of Wastes at Sea, which met in London in November 1972 at the invitation of the United Kingdom, adopted this instrument, generally known as the London Convention.

When the Convention came into force on 30 August 1975, IMO was made responsible for the Secretariat duties related to it.

The Convention has a global character, and contributes to the international control and prevention of marine pollution. It prohibits the dumping of certain hazardous materials, requires a prior special permit for the dumping of a number of other identified materials and a prior general permit for other wastes or matter.

"Dumping" has been defined as the deliberate disposal at sea of wastes or other matter from vessels, aircraft, platforms or other man-made structures, as well as the deliberate disposal of these vessels or platforms themselves.

Wastes derived from the exploration and exploitation of sea-bed mineral resources are, however, excluded from the definition. The provision of the Convention shall also not apply when it is necessary to secure the safety of human life or of vessels in cases of force majeure.

Among other requirements, Contracting Parties undertake to designate an authority to deal with permits, keep records, and monitor the condition of the sea.

Other articles are designed to promote regional co-operation, particularly in the fields of monitoring and scientific research.

Annexes list wastes which cannot be dumped and others for which a special dumping permit is required. The criteria governing the issuing of these permits are laid down in a third Annex which deals with the nature of the waste material, the characteristics of the dumping site, and method of disposal.

THE 1978 AMENDMENTS—INCINERATION

Adoption: 12 October 1978
Entry into force: 11 March 1979

The amendments affect Annex I of the Convention and are concerned with the incineration of wastes and other matter at sea.

THE 1978 AMENDMENTS—DISPUTES

Adoption: 12 October 1978
Entry into force: 60 days after being accepted by two thirds of Contracting
    Parties.
Status: see status of conventions

As these amendments affect the articles of the Convention they are not subject to the tacit acceptance procedure and will enter into force one year after being positively accepted by two thirds of Contracting Parties. They introduce new procedures for the settlement of disputes.

THE 1980 AMENDMENTS—LIST OF SUBSTANCES

Adoption: 24 September 1980
Entry into force: 11 March 1981

These amendments are related to those concerned with incineration and list substances which require special care when being incinerated.

THE 1989 AMENDMENTS

Adoption: 3 November 1989
Entry into force: 19 May 1990

The amendments qualify the procedures to be followed when issuing permits under Annex III. Before this is done, consideration has to be given to whether there is sufficient scientific information available to assess the impact of dumping.

THE 1993 AMENDMENTS

Adoption: 12 November 1993
Entry into force: 20 February 1994

The amendments banned the dumping into sea of low-level radioactive wastes. In addition, the amendments:

- phased out the dumping of industrial wastes by 31 December 1995
- banned the incineration at sea of industrial wastes

Although all three disposal methods were previously permitted under the Convention, attitudes towards the use of the sea as a site for disposal of wastes have changed over the years.

In 1983 the Contracting Parties to the LC adopted a resolution calling for a moratorium on the sea dumping of low-level radioactive wastes. Later resolutions called for the phasing-out of industrial waste dumping and an end to the incineration at sea of noxious liquid wastes.

1996 PROTOCOL

Adoption: 7 November 1996
Entry into force: 24 March 2006

The Protocol is intended to replace the 1972 Convention.

It represents a major change of approach to the question of how to regulate the use of the sea as a depository for waste materials.

One of the most important innovations is to introduce (in Article 3) what is known as the "precautionary approach." This requires that "appropriate preventative measures are taken when there is reason to believe that wastes or other matter introduced into the marine environment are likely to cause harm even when there is no conclusive evidence to prove a causal relation between inputs and their effects.

"The article also states that "the polluter should, in principle, bear the cost of pollution" and it emphasizes that Contracting Parties should ensure that the Protocol should not simply result in pollution being transferred from one part of the environment to another.

The 1972 Convention permits dumping to be carried out provided certain conditions are met. The severity of these conditions varies according to the danger to the environment presented by the materials themselves and there is a "black list" containing materials which may not be dumped at all.

The 1996 Protocol is much more restrictive.

PERMITTED DUMPING

Article 4 states that Contracting Parties "shall prohibit the dumping of any wastes or other matter with the exception of those listed in Annex 1."
    These are:

1. Dredged material
2. Sewage sludge
3. Fish waste, or material resulting from industrial fish processing operations
4. Vessels and platforms or other man-made structures at sea
5. Inert, inorganic geological material
6. Organic material of natural origin
7. Bulky items primarily comprising iron, steel, concrete and similar unharmful materials for which the concern is physical impact and limited to those circumstances where such wastes are generated at locations, such as small islands with isolated communities, having no practicable access to disposal options other than dumping.

The only exceptions to this are contained in Article 8 which permits dumping to be carried out "in cases of force majeure caused by stress of weather, or in any case which constitutes a danger to human life or a real threat to vessels . . ."

Incineration of wastes at sea was permitted under the 1972 Convention, but was later prohibited under amendments adopted in 1993. It is specifically prohibited by Article 5 of the 1996 Protocol.

In recent years concern has been expressed at the practice of exporting wastes which cannot be dumped at sea under the 1972 Convention to non-Contracting Parties.

Article 6 of the Protocol states that "Contracting Parties shall not allow the export of wastes or other matter to other countries for dumping or incineration at sea."

Article 9 requires Contracting Parties to designate an appropriate authority or authorities to issue permits in accordance with the Protocol.

The Protocol recognizes the importance of implementation and Article 11 details compliance procedures under which, no later than two years after the entry into force of the Protocol, the Meeting of Contracting Parties "shall establish those procedures and mechanisms necessary to assess and promote compliance . . ."

A key provision is the so-called transitional period (Article 26) which allows new Contracting Parties to phase in compliance with the convention over a period of five years. This provision is supported by extended technical assistance provisions.

IMO is made responsible for Secretariat duties in relation to the Protocol (as it is by the 1972 Convention). Other Articles contain procedures for settling disputes (Article 16) and amendments. Amendments to the Articles shall enter into force "on the 60th day after two-thirds of Contracting Parties shall have deposited an instrument of acceptance of the amendment with the Organization" (meaning IMO).

The Protocol contains three annexes. Annex 1 is described above and the other two deal with assessment of wastes and arbitral procedures.

Amendments to the annexes are adopted through a tacit acceptance procedure under which they will enter into force not later than 100 days after being adopted.

The amendments will bind all Contracting Parties except those which have explicitly expressed their non-acceptance.

2006 AMENDMENTS TO THE 1996 PROTOCOL

Adoption: 2 November 2006
Entry into force: 10 February 2007

Storage of carbon dioxide (CO_2) under the seabed will be allowed from 10 February 2007, under amendments to an international convention governing the dumping of wastes at sea.

Contracting Parties to the London Protocol, at their first meeting held in London from 30 October to 3 November, adopted amendments to the 1996 Protocol to the Convention on the Prevention of Marine Pollution by Dumping of Wastes and Other Matter, 1972 (London Convention). The amendments regulate the sequestration of CO_2 streams from CO_2 capture processes in sub-seabed geological formations.

Parties also agreed that guidance on the means by which sub-seabed geological sequestration of carbon dioxide can be conducted should be developed as soon as possible. This will, when finalized, form an important part of the regulation of this activity. Arrangements have been made to ensure that this guidance will be considered for adoption at the 2nd Meeting of Contracting Parties in November 2007.

This means that a basis has been created in international environmental law to regulate carbon capture and storage (CCS) in sub-seabed geological formations, for permanent isolation, as part of a suite of measures to tackle the challenge of climate change and ocean acidification, including, first and foremost, the need to further develop low carbon forms of energy. In practice, this option would apply to large point sources of CO_2 emissions, including power plants, steel and cement works.

The amendments, which will enter into force 100 days after adoption (i.e., on 10 February 2007), state that carbon dioxide streams may only be considered for dumping, if: disposal is into a sub-seabed geological formation; they consist overwhelmingly of carbon dioxide (they may contain incidental associated substances derived from the source material and the capture and sequestration processes used); and no wastes or other matter are added for the purpose of disposing of them.

Appendix C: Prediction of Sediment Toxicity Using Consensus Based Freshwater Sediment Quality Guidelines: USGS. 2000. Prediction of sediment toxicity using consensus based freshwater sediment quality guidelines. EPA 905/R-00/007, June 2000.

TABLE C.1
Sediment Quality Guidelines That Reflect Probable Effect Concentrations (PECs)

Substance	PEL	SEL	TET	ERM	PEL-HA28	Consensus-Based PEC
Metals (in mg/kg DW)						
Arsenic	17	33	17	85	48	33.0*
Cadmium	3.53	10	3	9	3.2	4.98*
Chromium	90	110	100	145	120	111*
Copper	197	110	86	390	100	149*
Lead	91.3	250	170	110	82	128*
Mercury	0.486	2	1	1.3	NG	1.06
Nickel	36	75	61	50	33	48.6*
Zinc	315	820	540	270	540	450*
Polycyclic Aromatic Hydrocarbons (in µg/kg DW)						
Anthracene	NG	3700	NG	960	170	845
Fluorene	NG	1600	NG	640	150	536
Naphtalene	NG	NG	600	2100	140	561*
Phenanthrene	515	9500	800	1380	410	1170*
Benz[a]anthracene	385	14800	500	1600	280	1050*
Benzo[a]pyrene	782	14400	700	2500	320	1450*
Chrysene	862	4600	800	2800	410	1290*
Fluoranthene	2355	10200	2000	3600	320	2230
Pyrene	875	8500	1000	2200	490	1520*
Total PAHs	NG	100000	NG	35000	3400	22800*
Polychlorinated Biphenyls (PCBs) (in µg/kg DW)						
Total PCBs	**277**	**5300**	**1000**	**400**	**240**	676*
Organochlorine Pesticides (in µg/kg DW)						
Chlordane	8.9	60	30	6	NG	17.6
Dieldrin	6.67	910	300	8	NG	61.8
Sum DDD	8.51	60	60	20	NG	28.0
Sum DDE	6.75	190	50	15	NG	31.3*
Sum DDT	NG	710	50	7	NG	62.9
Total DDTs	4450	120	NG	350	NG	572
Endrin	62.4	1300	500	45	NG	207
Heptachlor epoxide	2.74	50	30	NG	NG	16.0
Lindane	1.38	10	9	NG	NG	4.99

PECs: concentrations above which harmful effects are likely to be observed (MacDonald et al., 2000). An "*" designates a reliable PEC (>20 samples and >75% correct classification as toxic).
PEL: probable effect level, dry weight (Smith et al., 1996)
SEL: severe effect level, dry weight (Persaud et al., 1993)
TET: Toxic effect threshold, dry weight (EC & MENVIQ, 1992)
ERM: Effects range median; dry weight (Long and Morgan, 1991)
PEL-HA28: probable effect level for *Hyalella azteca*; 28-day test; dry weight (USEPA, 1996)
NG: No guideline

REFERENCES

Environment Canada and Ministere de l'Environnement du Quebec (EC and MENVIQ). 1992. Interim criteria for quality assessment of St. Lawrence River sediment. ISBN 0-662-19849-2, Environment Canada, Ottawa, ON.

Long, E.R. and Morgan, L.G. 1991. The potential for biological effects of sediment sorbed contaminants in the National Status and Trends Program. NOAA Technical Memorandum NOS OMA 52. National Oceanic and Atmospheric Administration, Seattle, WA, 175 pp. + Appendices.

MacDonald, D.D., Ingersoll, C.G., and Berger, T. 2000. Development and evaluation of consensus-based sediment quality guidelines for freshwater ecosystems. *Arch. Environ. Contam. Toxicol.* 39: 20–31.

Persaud, D., Jaagumagi, R., and Hayton, A. 1993. Guidelines for the protection and management of aquatic sediment quality in Ontario. Water Resources Branch, Ontario Ministry of the Environment, Toronto, ON, 27 pp.

Smith, S.L., Macdonald, D.D., Keenelysied, K.A., Ingersoll, C.G., and Field, J. 1996. A preliminary evaluation of sediment quality assessment values for freshwater ecosystems. *J. Great Lakes Res.* 22: 624–638.

USEPA. 1996. Calculation and evaluation of sediment effect concentrations for the amphipod *Hyalella azteca* and the midge *Chironomus riparius*. EPA 905/R-96/008, Chicago, IL.

Appendix D: International Sediment Quality Criteria

HONG KONG

Hong Kong classifies sediments based on their contaminant levels with reference to the Chemical Exceedance Levels (CEL) shown below (Hong Kong Environment and Transport Bureau (ETWB 34/2002), http://www.devb-wb.gov.hk/UtilManager/tc/2002/C-2002-34-0-1.pdf).

Sediment Quality Criteria for the Classification of Sediments in Hong Kong

Contaminants	Lower Chemical Exceedance Level (LCEL)	Upper Chemical Exceedance Level (UCEL)
Metals (mg/kg dry wt.)		
Cadmium (Cd)	1.5	4
Chromium (Cr)	80	160
Copper (Cu)	65	110
Mercury (Hg)	0.5	1
Nickel (Ni)*	40	40
Lead (Pb)	75	110
Silver (Ag)	1	2
Zinc (Zn)	200	270
Metalloid (mg/kg dry wt.)		
Arsenic (As)	12	42
Organic-PAHs (µg/kg dry wt.)		
Low Molecular Weight PAHs	550	3160
High Molecular Weight PAHs	1700	9600
Organic-non-PAHs (µg/kg dry wt.)		
Total PCBs	23	180
Organometallics (µg TBT/L in Interstitial water)		
Tributyltin*	0.15	0.15

The contaminant level is considered to have exceeded the UCEL if it is greater than the value shown. The sediment is classified into three categories based on its contaminant levels:

- *Category L*: sediment with all contaminant levels not exceeding the Lower Chemical Exceedance Level (LCEL). The material must be dredged, transported, and disposed of in a manner which minimizes the loss of contaminants either into solution or by resuspension.
- *Category M*: sediment with any one or more contaminant levels Lower Chemical Exceedance Level (LCEL) and none exceeding Upper Chemical Exceedance Level (UCEL). The material must be dredged and transported with care and must be effectively isolated from the environment upon final disposal unless appropriate biological tests demonstrate that the material will not adversely affect the marine environment.
- *Category H*: Sediment with any one or more contaminant levels exceeding Upper Chemical Exceedance Level (UCEL). The material must be dredged and transported with great care and must be effectively isolated from the environment upon final disposal.

THE REPUBLIC OF KOREA

Action List for the Degraded Material Disposal at Sea in the Republic of Korea

Parameter	1st Level (Upper Level) mg/kg Dry Weight	2nd Level (Lower Level) mg/kg Dry Weight
Chromium and its compounds	370	80
Zinc and its compounds	410	200
Copper and its compounds	270	65
Cadmium and its compounds	10	2.5
Mercury and its compounds	1.2	0.3
Arsenic and its compounds	70	20
Lead and its compounds	220	50
Nickel and its compounds	52	35
Total polychlorinated biphenyls	0.180	0.023
Total polyaromatic hydrocarbons	45	4

Notes: Total polychlorinated biphenyls is the sum of contents of PCB-28, PCB-101, PCB-138, PCB-153, and PCB-180 congeners in a sample. Total polyaromatic hydrocarbons is sum content of naphthalene, phenanthrene, anthracene, benzo(a)pyrene, benzo(a) anthracene, fluoranthene, benzo(b), and fluoranthene in a sample.

AUSTRALIA AND NEW ZEALAND

AUSTRALIA AND NEW ZEALAND GUIDELINES FOR FRESH AND MARINE WATER QUALITY

National Ocean Disposal Guidelines for Dredged Material

Environment Australia, May 2002, ISBN 0 6425 4831 5. http://www.environment.gov.au/coasts/pollution/dumping/guidelines/pubs/guidelines.pdf.

CANADA

Canadian Council of the Environment. 2001. Canadian sediment Quality Guidelines for the protection of aquatic life. Updated. Environmental quality guidelines, 1999. Canadian Council of Ministers of the Environment, Winnipeg, Canada.

CCME. 1999. Protocol for the Derivation of Canadian Sediment Quality Guidelines for the Protection of Aquatic Life, CCME EPC98E. http://www.ccme.ca/assets/pdf/sedqg_protocol.pdf.

Canadian Disposal at Sea Program website. http://www.ec.gc.ca/seadisposal/reports/index_e.htm#sqg.

UNITED STATES

USEPA

National Sediment Inventory. 2004. Appendix C—Screening Values for Chemicals Evaluated, http://www.epa.gov/waterscience/cs/report/2004/nsqs2ed-complete.pdf#page=213.

NOAA. 1999. Sediment Quality Guidelines developed for the National Status and Trends Program, National Oceanic and Atmospheric Administration, http://response.restoration.noaa.gov/book_shelf/121_sedi_qual_guide.pdf.

USEPA. 1997. Ecological Risk Assessment Guidance for Superfund: Process for Designing and Conducting Ecological Risk Assessments. EPA 540-R-97-006, 1997. U.S. Environmental Protection Agency, Office of Solid Waste and Emergency Response, Washington, DC, http://www.epa.gov/oswer/riskassessment/ecorisk/ecorisk.htm.

USACE

USEPA/USACE. 1991. Evaluation of Dredged Material Proposed for Ocean Disposal. Testing Manual, EPA 503/8-91/001. U.S. Environmental Protection Agency and U.S. Army Corps of Engineers, Washington, DC.

USEPA/USACE. 1998. Evaluation of dredged material proposed for discharge in waters of the U.S. testing manual. EPA-823-B-98-004, Washington, DC, http://el.erdc.usace.army.mil/elmodels/pdf/inlandb.pdf.

USGS. 2002. Prediction of Sediment Toxicity using consensus based freshwater sediment quality guidelines. EPA/905/R-00/007, June 2000, United States Geological Survey (USGS), http://www.cerc.usgs.gov/pubs/center/pdfdocs/91126.pdf.

U.S. STATE GUIDELINES

Florida

1994 Florida Sediment Quality Assessment Guidelines (SQAGs)

Chemical Parameter	Sediment Quality Assessment Guidelines TEL mg/kg Dry Weight (Parts per Million (ppm) Dry)	Sediment Quality Assessment Guidelines PEL mg/kg Dry Weight (Parts per Million (ppm) Dry)
Arsenic	7.24	41.6
Cadmium	0.676	4.21
Chromium	52.3	160
Copper	18.7	108
Lead	30.2	112
Mercury	0.13	0.696
Silver	0.733	1.77
Zinc	124	271
	μg/kg	μg/kg
Naphthalene	34.6	391
Acenaphthylene	5.87	128
Acenaphthene	6.71	88.9
Fluorene	21.2	144
Phenanthrene	86.7	544
Anthracene	46.9	245

1994 Florida Sediment Quality Assessment Guidelines (SQAGs)

Chemical Parameter	Sediment Quality Assessment Guidelines TEL mg/kg Dry Weight (Parts per Million (ppm) Dry)	Sediment Quality Assessment Guidelines PEL mg/kg Dry Weight (Parts per Million (ppm) Dry)
2-Methylnaphthalene	20.2	201
Total lmw-PAHs	312	1442
Fluoranthene	113	1494
Pyrene	1,000	1400
Benz(a)anthracene	74.8	693
Chrysene	108	846
Benzo(a)pyrene	88.8	763
Indeno(1,2,3,-c,d)pyrene	34	88
Dibenzo(a,h)anthracene	113	1494
Pesticides		
Chlordane	2.26	4.79
p,p'-DDD	1.22	7.81
*p,p'*DDE	2.07	374
p,p'-DDT	1.19	4.77
Total DDT	3.89	51.7
Dieldrin	0.715	4.3
Lindane	0.32	0.99
Bis(2-ethylhexyl)phthalate	182	2647
Total PCBs	21.6	189

Source: http://www.dep.state.fl.us/waste/quick_topics/publications/pages/default.htm.
Note: TEL= toxic effect level; PEL= probable effect level.

New York

New York State Department of Environmental Conservation, Division of Fish, Wildlife and Marine Resources. 1999. Technical Guidance for Screening Contaminated Sediments, Jan. 1999, http://www.dec.ny.gov/docs/wildlife_pdf/seddoc.pdf.

Washington State

Sediment Quality Chemical Criteria

The Sediment Management Standards currently contain two sets of numeric chemical criteria that apply to Puget Sound marine sediments:

1. The "no effects" level—the Sediment Quality Standards, WAC 172-204-320—used as a sediment quality goal for Washington State sediments (shown below), and
2. The "minor adverse effects" level—The Sediment Impact Zone Maximum Level, WAC 173-204-420; and the Sediment Cleanup Screening Level/Minimum Cleanup Level, WAC 173-204-520—used as an upper regulatory level for source control and cleanup decision making (shown below).

To understand the context in which the criteria are used, see the Sediment Management Standards regulation.

Chemical Parameter	Sediment Quality Standards WAC 173-204-320 (a)	Sediment Impact Zone Maximum Level, WAC 173-204-420 (a); and Sediment Cleanup Screening Level/Minimum Cleanup Level, WAC 173-204-520 (a)
	mg/kg Dry Weight (Parts per Million (ppm) Dry)	mg/kg Dry Weight (Parts Per Million (ppm) Dry)
Arsenic	57	93
Cadmium	5.1	6.7
Chromium	260	270
Copper	390	390
Lead	450	530
Mercury	0.41	0.59
Silver	6.1	6.1
Zinc	410	960
	mg/kg Organic Carbon (C) (ppm Carbon)	mg/kg Organic Carbon (C) (ppm Carbon)
LPAH (b,d)	370	780
Naphthalene	99	170
Acenaphthylene	66	66
Acenaphthene	16	57
Fluorene	23	79
Phenanthrene	100	480
Anthracene	220	1200
2-Methylnaphthalene	38	64
HPAH (b,e)	960	5300
Fluoranthene	160	1200
Pyrene	1,000	1400
Benz(a)anthracene	110	270
Chrysene	110	460
Total Benzofluoranthenes (b,f)	230	450
Benzo(a)pyrene	99	210

Indeno(1,2,3,-c,d)pyrene	34	88
Dibenzo(a,h)anthracene	12	33
BENZO(G,H,I)PERYLENE	31	78
1,2-Dichlorobenzene	2.3	2.3
1,4-Dichlorobenzene	3.1	9
1,2,4-Trichlorobenzene	0.81	1.8
Hexachlorobenzene	0.38	2.3
Dimethyl Phthalate	53	53
Diethyl Phthalate	61	110
Di-n-butyl Phthalate	220	1700
Butyl Benzyl Phthalate	4.9	64
Bis(2-ethylhexyl) Phthalate	47	78
Di-n-octyl Phthalate	58	4500
Dibenzofuran	15	58
Hexachlorobutadiene	3.9	6.2
n-Nitrosodiphenylamine	11	11
Total PCBs (b)	12	65

	µg/kg Dry Weight (Parts per Billion (ppb) Dry)	µg/kg Dry Weight (Parts per Billion (ppb) Dry)
Phenol	420	1200
2-Methylphenol	63	63
4-Methylphenol	670	670
2,4-Dimethyl Phenol	29	29
Pentachlorophenol	360	690
Benzyl Alcohol	57	73
Benzoic Acid	650	650

Source: Washington State Department of Ecology Toxic Cleanup Program. 2008. Sediment Quality Chemical Criteria. http://www.ecy.wa.gov/programs/tcp/smu/sed_chem.htm.

Note:
  a. Where laboratory analysis indicates a chemical is not detected in a sediment sample, the detection limit shall be reported and shall be at or below the Marine Sediment Quality Standards chemical criteria value set in this table.
  b. Where chemical criteria in this table represent the sum of individual compounds or isomers, the following methods shall be applied:
      i. Where chemical analyses identify an undetected value for every individual compound/ isomer then the single highest detection limit shall represent the sum of the respective compounds/isomers; and
      ii. Where chemical analyses detect one or more individual compound/isomers, only the detected concentrations will be added to represent the group sum.
  c. The listed chemical parameter criteria represent concentrations in parts per million, "normalized," or expressed, on a total organic carbon basis. To normalize to total organic carbon, the dry weight concentration for each parameter is divided by the decimal fraction representing the percent total organic carbon content of the sediment.
  d. The LPAH criterion represents the sum of the following "low molecular weight polynuclear aromatic hydrocarbon" compounds: Naphthalene, Acenaphthylene, Acenaphthene, Fluorene, Phenanthrene, and Anthracene. The LPAH criterion is not the sum of the criteria values for the individual LPAH compounds as listed.
  e. The HPAH criterion represents the sum of the following "high molecular weight polynuclear aromatic hydrocarbon" compounds: Fluoranthene, Pyrene, Benz(a)anthracene, Chrysene, Total Benzofluoranthenes, Benzo(a)pyrene, Indeno(1,2,3,-c,d)pyrene, Dibenzo(a,h)anthracene, and Benzo(g,h,i)perylene. The HPAH criterion is not the sum of the criteria values for the individual HPAH compounds as listed.
  f. The TOTAL BENZOFLUORANTHENES criterion represents the sum of the concentrations of the "B," "J," and "K" isomers.

WISCONSIN

State of Wisconsin, Department of Natural Resources, Consensus Sediment Quality Guidelines, Recommendations for Use and Application, Interim Guidance, Contaminated Sediment Standing Team. WT-732 2003. Dec. 2003. http://dnr.wi.gov/org/aw/rr/technical/cbsqg_interim_final.pdf.

EUROPE

EUROPEAN LEGISLATION

EC Legislation

Several western European countries have developed their own placement policies or guidelines, but certain EC Directives govern the Placement and/or use of dredged material in EC countries under the definition of "waste." This section reviews both the EC Directives and the individual countries' policies.

Classification of Dredged Material in the EC Region

Several EU Member States have defined or proposed sediment quality levels that trigger various levels of action. While definitions vary, they may be generalized as:

Class 1—Below action Level 1: sea Placement permitted

Class 2—Between Action Levels 1 and 2: sea placement permitted with restrictions (e.g., monitoring)

Class 3—Higher than Action Level 2: sea placement permitted only under very specific conditions

Here are some of the individual states.

BELGIUM

Sediment Quality Criteria for Belgium, on Metals and Organics in Dredged Material

Parameter	Action Level 1 (Target Value) (ppm D.M.)	Action Level 2 (Limit Value) (ppm D.M.)
Hg	0.3	1.5
Cd	2.5	7
Pbd	70	350
Zn	160	500
Ni	70	280
As	20	100
Cr	60	220
Cu	20	100
TBT	3	7
Mineral oil	14 mg/goc	36 mg/goc
PAHs	70 µg/goc	180 µg/goc
PCBs	2 µg/goc	2 µg/goc

GOC, gram organic carbon

FINLAND

The action levels for dredged material in Finland were adopted by the ministry of the Environment on 19 May 2004. These values are still, however, guidance values and not binding forms. The aim is to be able to give binding norms within a few years. All measured contaminant contents are normalized to a "standard soil" composition (10% organic material and 25% clay). The values in the table refer to the normalized values.

Contaminant	Action Level 1 (ppm Dry Weight)	Action Level 2 (ppm Dry Weight)
Hg	0.1	1
Cd	0.5	2.5
Cr	65	70
Cu	50	290
Pb	40	200
Ni	45	60
Zn	170	500
As	15	60
PAHs		
Naphthalene	0.01	0.1
Anthracene	0.01	0.1
Phenanthrene	0.05	0.5
Fluoranthene	0.3	3
Benzo[a]anthracene	0.03	0.4
Chrysene	1.1	11
Benzo(k)fluoranthene	0.2	2
Benzo[a]pyrene	0.3	3
Benzo[ghi]perylene	0.8	8
Indeno(123-cd)pyrene	0.6	6
Mineral oil	500	1500
DDT+DDE+DDD	0.01	0.03
	ppb Dry Weight	**ppb Dry Weight**
PCB (IUPAC-numbers)		
28	1	30
52	1	30
101	4	30
118	4	30
138	4	30
153	4	30
180	4	30
Tributyltin (TBT)	3	200
	ng WHO-TEQ/kg	**ng WHO-TEQ/kg**
Dioxins and furans (PCDD and PCDF)	20	50

FRANCE

If analysis shows that concentrations are less than action level 1, a general permit is given without specific study.

If analysis shows that concentrations exceed action level 2, dumping at sea may be prohibited, especially when this dumping does not constitute the least detrimental solution for the environment (particularly with respect to the other solutions, in situ, or on land). These values do not consider the toxic character and bioavailability of each element.

If analysis shows that concentrations are situated between action level 1 and action level 2, a more comprehensive study might be necessary. The content of these studies will be established on a case-by-case basis, taking account of the local circumstances and the sensitivity of the environment.

The action levels are shown in the following table.

Substances	Action Level 1 (ppm Dry Weight)	Action Level 2 (ppm Dry Weight)	Substance	Action Level 1 (ppm Dry Weight)	Action Level 2 (ppm Dry Weight)
Metals			PCB		
Hg	0.4	0.8	CB 28	0.025	0.05
Cd	1.2	2.4	52	0.025	0.05
As	25	50	101	0.050	0.05
Pb	100	200	118	0.025	0.10
Cr	90	180	180	0.025	0.05
Cu	45	90	138	0.05	0.10
Zn	276	552	153	0.05	0.10
Ni	37	74	Total PCBs	0.5	1.0

GERMANY

Sediment Quality for the German Federal Waters and Navigation Administration on Trace Metals and Organic Contaminants in Dredged Material (Sediment Fraction<20 μm)

		Action Level 1	Action Level 2
As	ppm	30	150
Ca	ppm	2.5	12.5
Cr	ppm	150	750
Cu	ppm	40	200
Hg	ppm	1	5
Ni	ppm	50	250
Pb	ppm	100	500

Sediment Quality for the German Federal Waters and Navigation Administration on Trace Metals and Organic Contaminants in Dredged Material (Sediment Fraction<20 µm)

		Action Level 1	Action Level 2
Zn	ppm	350	1750
PCB28	ppb	2	6
PCB52	ppb	1	3
PCB101	ppb	2	6
PCB118	ppb	3	10
PCB138	ppb	4	12
PCB153	ppb	5	15
PCB180	ppb	2	6
Sum of 7 PCBs	ppb	20	60
α-Chlorcyclohexane	ppb	0.4	1
γ-Chlorcyclohexane	ppb	0.2	0.6
Hexachlorobenzene	ppb	2	6
Pentachlorobenzene	ppb	1	3
p,p'-DDT	ppb	1	3
p,p'-DDE	ppb	1	3
p,p'-DDD	ppb	3	10
PAH* (sum of 6 PAH)	ppm	1	3
Hydrocarbons	ppm	300	1000

*   *Total of 6 PAH compounds:* fluoranthene, benzo(b)fluoranthene, benzo(k)fluoranthene, benzo[a]pyrene, benzo[ghi]perylene, and indeno(1,2,3-cd)pyrene.

Action Levels for the German Federal Waters and Navigation Administration on Tributyltin (TBT) in Dredged Material

Action Level 1	Action Level 2	Unit	Valid from
20	600	µg TBT/kg total sediment	2001
20	300	µg TBT/kg total sediment	2005
20	60	µg TBT/kg total sediment	2010

IRELAND

Guidelines for assessment of dredged material for placement in Irish waters have been published.

Provisional Irish Action Levels in mg kg^{-1} Dry Weight

Chemical	Category 1	Category 2	Category 3
As	<10	10–80	>80
Cd	<1	1-3	>3
Cr	<100	100–300	>300
Cu	<50	50–200	>200
Hg	<0.3	0.3–5.0	>5
Ni	<50	50–200	>200
Pb	<50	50–400	>400
Zn	<400	400–700	>700
PCB (7)	<0.01	0.01–0.1	>0.1
TBT	<0.1	0.1–0.5	>0.5
Total PCB	<0.1	0.1-1.0	>1.0

THE NETHERLANDS

The Dutch classification system for dredging material has recently been revised:

- *Target value*: Indicates the level below which risks to environment are considered to be negligible, at the present state of knowledge.
- *Limit value*: Concentration at which the water sediment is considered as relatively clean. The limit value is objective for the year 2000.
- *Reference value*: A reference level indicating whether dredged sediment is still fit for discharge in surface water, under certain conditions, or should be treated otherwise. It indicates the maximum allowable level above which the risks for the environment are unacceptable.
- *Intervention value*: An inactive value, indicating that remediation may be urgent, owing to increased risks to public health and the environment.
- *Signal value*: Only for heavy metals. Concentration level of heavy metals above which the need for cleaning up should be investigated.

Constants in the Correction of Measured Levels for Heavy Metals and Arsenic Based on the Local Sediment Composition (Derived from Reference Value)

Metal	A	B	C
Zn	50	3	1.5
Cu	15	0.5	0.6
Cr	50	2	0
Pb	50	1	1
Cd	0.4	0.0007	0.021
Ni	10	1	0
Hg	0.2	0.0034	0.0017
As	15	0.4	0.4

The water sediment standards now existing have been based upon information which estimates the effects on the aquatic ecosystem. In addition, the water sediment composition influences the standards. For the availability of heavy metals and arsenic, clay fraction (lute particle size <2 µm) and the quantity of organic material are of particular importance. For the availability of organic compound, the organic substance level is a determining factor. The standards are set for sediment containing 25% of lute and 10% of organic substance. Conversion toward the standard sediment composition is done in conformity to the method followed by the WOB (Water Sediment Study Group), which is also applied to calculate the reference values for soil quality.

Classification of water sediment:
Class 0 is below target value and can be spread over the land without restrictions.
Class 1 exceeds the target value, but it is below the limit value and is allowed to be disposed unless the soil quality is not significantly impaired.
Class 2 does not meet the limit value, but is below the reference value and can be spread in surface water or on land, under certain conditions.
Class 3 does not meet the reference value, but remains below the intervention value, and should be stored under controlled conditions; specific requirements can be set, depending on the storage location.
Class 4 does not meet the intervention value, and should be contained in isolation in deep pits or on land, in order to minimize the influence on the surroundings.

Target and Other Values

Parameter	Unit	Target Value	Limit Value	Reference Value	Intervention Value	Signal Value
Arsenic	mg/kg ds	29	55	55	55	150
Cadmium	mg/kg ds	0.8	2	7.5	12	30
Chromium	mg/kg ds	100	380	380	380	1000
Copper	mg/kg ds	35	35	90	190	400
Mercury	mg/kg ds	0.3	0.5	1.6	10	15
Lead	mg/kg ds	85	530	530	530	1000
Nickel	mg/kg ds	35	35	45	210	200
Zinc	mg/kg ds	140	480	720	720	2500
PAH total 10 PAH*	mg/kg ds	1	1	10	40	—
PCB-28	µg/kg ds	1	4	730	—	—
PCB-52	µg/kg ds	1	4	730	—	—
PCB-101	µg/kg ds	4	4	730	—	—
PCB-118	µg/kg ds	4	4	730	—	—
PCB-138	µg/kg ds	4	4	730	—	—
PCB-153	µg/kg ds	4	4	730	—	—
PCB-180	µg/kg ds	4	4	730	—	—
Total 6 PCB	µg/kg ds	20/0	—	—	—	—
Total 7 PCB	µg/kg ds	—	—	200	1000	—
Chlordane	µg/kg ds	10	20	—	—	—
α-HCH	µg/kg ds	2.5	—	20	—	—
β-HCH	µg/kg ds	1	—	20	—	—
γ-HCH (lindane)	µg/kg ds	0.05	1	20	—	—
HCH-compounds	µg/kg ds	—	—	—	2000	—
Heptachlor	µg/kg ds	2.5	—	—	—	—
Heptachlorepoxide	µg/kg ds	2.5	—	—	—	—
Heptachlor+epoxide	µg/kg ds	—	20	20	—	—
Aldrin	µg/kg ds	2.5	—	—	—	—
Dieldrin	µg/kg ds	0.5	20	—	—	—
Total aldrin & dieldrin	µg/kg ds	—	40	40	—	—
Endrin	µg/kg ds	1	40	40	—	—
Drins	µg/kg ds	—	—	—	4000	—
DDT (incl. DDD & DDE)	µg/kg ds	2.5	10	20	4000	—
α-Endosulfan	µg/kg ds	2.5	—	—	—	—
α-Endosulfan + sulphate	µg/kg ds	—	10	20	—	—
Hexachlorobutadiene	µg/kg ds	2.5	—	100	—	—
Total pesticides	µg/kg ds					
Pentachlorobenzene	µg/kg ds	2.5	300	300	—	—
Hexachlorobenzene	µg/kg ds	2.5	4	20	—	—
Pentachlorophenol	µg/kg ds	2	20	5000	5000	—

Mineral oil	mg/kg ds	50	100	3000	5000	—
EOX	mg/kg ds	—	0	7	—	—

*   *Naphthalene, benzo(a)anthracene, benzo(b)fluoranthene, benzo(k)fluoranthene, benzo[a]pyrene, phenanthrene, benzo[ghi]perylene, indeno(1,2,3-cd)pyrene, anthracene, chrythene, fluoranthene.*

Dredged Material Standards for the Netherlands

ppm Dry Weight	Action Level 1[a]	Action Level 2[b]
As	29	29
Cd	0.8	4
Cr	100	120
Cu	36	60
Hg	0.3	1.2
Pb	85	110
Ni	35	45
Zn	140	365
Mineral oil (C10–40)	50	1250
Sum 10 PAHs[c]		8
Sum 7 PCBs[d]		0.1
α-CHC	0.003	—
β-CHC	0.009	—
γ-CHC (lindane)	0.00005	0.02
Sum HCHs	0.01	—
Heptachlor	0.007	—
Heptachlorepoxide	0.0000002	0.02
Aldrin	0.00006	0.03
Dieldrin	0.0005	0.03
Endrin	0.00004	0.03
Sum Aldrin + Dieldrin + Endrin	0.005	—
DDT	0.00009	—
DDD	0.00002	—
DDE	0.00001	—
Sum DDT+ DDD+DDE	0.01	0.02
Hexachlorbenzene	0.00005	0.02
TBT	0.000007	0.24(100 µg Sn/kg dw)
Sum organic compounds	0.001	

[a] General environmental quality objective (water system).
[b] Numerical values for the content test distribution into salt waters (2001).
[c] *Naphthalene, benzo(a)anthracene, benzo(b)fluoranthene, benzo(k)fluoranthene, benzo[a]pyrene, phenanthrene, benzo[ghi]perylene, indenopyrene, anthracene, chrysene, fluoranthene.
[d] PCBs 28, 52, 101, 118, 138, 153, and 180.

NORWAY

The Norwegian sediment criteria for Classification of Environmental Quality and Degree of Pollution (CEQDP) in fjords and coastal waters represent the basis for managing dredging and dredged material.

Dredged Material Standards for Norway

Parameter	Category1 Good/Fair (Class I & II)	Category 2 Poor/Bad (Class III & IV)	Category 3 Very Bad (Class V)
Metals (ppm Dry Weight)			
Arsenic	<20–80	80–1000	>1000
Lead	<30–120	120–1500	>1500
Fluoride	<800–3000	3000–20000	>20000
Cadmium	<0.25–1	1–10	>10
Copper	<35–150	150–1500	>1500
Mercury	<0.15–0.6	0.6–5	>5
Chromium	<70–300	300–500	>5000
Nickel	<30–130	130–1500	1500
Zinc	<150–700	700–10000	10000
Silver	<0.3–1.3	1.3–10	>10
Organic Component (ppb Dry Weight)			
Sum PAH (EPA 16)	<300–2000	2000–20000	>20000
Benzo(a)pyrene	<10–50	50–500	>500
Sum PCB	<5–25	25–300	>300
Hexachlorobenzene	<0.5–2.5	2.5–50	>50
EPOCL[e]	<100–500	500–15000	>15000
2, 3, 7, 8-TCDD eqv.[f]	<0.3–0.12	0.12–1.5	>1.5

[e] Extractable persistent organic chloride.

[f] Total toxicity potential for polychlorinated dibenzofurans/dioxins, given as equivalents of the most toxic of these components (2,3,7,8-tetrachlorodibenzo-p-dioxin).

PORTUGAL

Dredged Material Classification for Portugal

Substance	Class 1	Class 2	Class 3	Class 4	Class 5
As	<20	50	100	500	>500
Cd	<1	30	5	10	>10
Cr	<50	100	400	1000	>1000
Cu	<35	150	300	500	>500
Hg	<0.5	1.5	3	10	>10
Pb	<50	150	500	1000	>1000
Ni	<30	75	125	250	>250
Zn	<100	600	1500	5000	>5000
PCB sum	<5	25	100	300	>300
PAH sum	<300	2000	6000	20000	>20000
HCB	<0.5	2.5	10	50	>50
Description	clean	vestiges of contamination	slightly contaminated	contaminated	very contaminated
Fate	aquatic environment and beaches	aquatic environment	aquatic environment with monitoring	landfill with special monitoring	landfill (residues have special treatment)

Notes:

1. Concentrations are upper bounds for each class.
2. Concentrations of metals are in mg/kg dry solids (ppm).
3. Concentrations of organics are in micrograms/kg dry solids (ppb).

SPAIN

Sediment Quality Criteria Applicable to Spanish Harbors

ppm Dry Weight	Action Level 1	Action Level 2
Hg	0.6	3
Cd	1	5
Pb	120	600
Cu	100	400
Zn	500	3000
Cr	200	1000
As	80	200
Ni	100	400
Sum 7 PCBs	0.03	0.1

SWEDEN

In Sweden, action levels are based on the following background concentrations. Information is not provided on the possible link between these conentrations and action levels:

Substance	Background Value (ppm Dry Weight)
As	10
Pb	10
Fe	40000
Cd	0.3
Co	15
Cu	20
Cr	20
Hg	0.1
Ni	15
Sn	1
V	20
Zn	125

THE UNITED KINGDOM

Most dredged material in the UK is placed at sea and is governed by part II of the Food and Environment Protection Act 1985 (FEPA).

Sediment Quality Criteria for the U.K. on Metals and Organics in Dredged Material

Contaminant	Existing Action Level 1 (mg·kg⁻¹) (ppm)	Existing Action Level 2 (mg·kg⁻¹) (ppm)	Suggested Revised Action Level 1 (mg·kg⁻¹) (ppm) Dry Weight	Suggested Revised Action Level 2 (mg·kg⁻¹) (ppm) Dry Weight
Arsenic(As)	20	50–100	20	70
Cadmium(Cd)	0.4	2	0.4	4
Chromium(Cr)	40	400	50	370
Copper(Cu)	40	400	30	300
Mercury(Hg)	0.3	3	0.25	1.5
Nickel(Ni)	20	200	50	150
Lead(Pb)	50	500	30	400
Zinc(Zn)	130	800	130	600
Tributyltin (TBT, DBT, MBT)	0.1	1	0.1	0.5

Sediment Quality Criteria for the U.K. on Metals and Organics in Dredged Material

Contaminant	Existing Action Level 1 (mg·kg⁻¹) (ppm)	Existing Action Level 2 (mg·kg⁻¹) (ppm)	Suggested Revised Action Level 1 (mg·kg⁻¹) (ppm) Dry Weight	Suggested Revised Action Level 2 (mg·kg⁻¹) (ppm) Dry Weight
Polychlorinated Biphenyls (PCBs)	0.02	0.2	0.02	0.18
Polyaromatic Hydrocarbons (PAHs)				
Acenaphthene			0.1	
Acenaphthylene			0.1	
Anthracene			0.1	
Fluorene			0.1	
Naphthalene			0.1	
Phenanthrene			0.1	
Benzo[a] anthracene			0.1	
Benzo[a] fluoranthene			0.1	
Benzo[k] fluoranthene			0.1	
Benzo[g]perylene			0.1	
Benzo[a]pyrene			0.1	
Benzo[g,h,i] perylene			0.1	
Dibenzo[a,h] anthracene			0.01	
Chrysene			0.1	
Fluoranthene			0.1	
Pyrene			0.1	
Indeno[1,2,3cd] pyrene			0.1	
Total hydrocarbons	100		100	
Booster biocide and brominated flame retardants[a]	—	—	—	—

[a] Provisional Action Levels for these compounds are subject to further investigation.

MID EAST

QATAR

The following information has been obtained from The Enviromental Guidelines and Enviromental Protection Criteria for Ras Laffan Industrial City. New regulations, permits, or standards are issued by an appropriate regulatory authority such as the Supreme Council for the Enviroment and Natural Reserves (SCENR).

Maximum Concentration of Containmants for Toxicity Characteristic

Contaminant	Regulatory Level mg/L	Contaminant	Regulatory Level mg/L
Arsenic	5.0	Hexachlorobenzene	0.13
Barium	100.0	Hexachlorobutadiene	0.5
Benzene	0.5	Hexachloroethane	3.0
Cadmium	1.0	Lead	5.0
Carbon tetrachloride	0.5	Lindane	0.4
Chlordane	0.03	Mercury	0.2
Chlorobenzene	100.0	Methoxychlor	10.0
Chloroform	6.0	Methyl ethyl ketone	200.0
Chromium	5.0	Nitrobenzene	2.0
o-Cresol	200.0	Pentrachlorophenol	100.0
m-Cresol	200.0	Pyridine	5.0
p-Cresol	200.0	Selenium	1.0
Cresol	200.0	Silver	5.0
Dichlorobenzene	7.5	Tetrachloroethylene	0.7
Dichloroethane	0.5	Toxaphene	0.5
Dichloroethylene	0.7	Trichloroethylene	0.5
Dinitrotoluene	0.13	Trichlorophenol	400.0
Endrin	0.02	Silvex	1.0
Heptachlor(and its epoxide)	0.008	Vinyl chloride	0.2

Index

A

Aberdeen Proving Ground, Maryland, 155
Acid leaching, 174–175
Acid volatile sulfide-simultaneously extracted metals (AVS:SEM), 91
Active capping, 144
Advection, cap thickness for, 140–144
Aeration, 145, *151*
Aerobic bioremediation, 155
Agriculture, 6–7, 31
Algae, 55, 90
Ammonium acetate, 48
Amphiboles, 23
Anacostia River, Maryland, 144
Analysis and evaluation, sediment
    chemical sediment quality, 82–92
    generic framework, 215–216
    mechanical properties, 77–79
    monitoring plans, 226–233
    natural recovery, 113–119
    physical properties, 79–82
Andersen, O., 61
Anjaneyulu, Y., 194
Apitz, S. E., 14
Appelo, C. A. J., 63
Aquablock™, 144
Aquatic geoenvironment
    agriculture and, 6–7
    heavy metals in, 5, 8
    mining industry and, 7
    presence of sediments in, 1–2
    sources of pollutants, 3–9
    sustainable development and, 3
    uses of sediments and water in, 31–32
Arias, Y. M., 127
Arsenic, 43–44
Atmospheric pollutants, 8
Attenuation. *See* Natural recovery (NR)
Australia, 279
Azcue, J. M., 136

B

Baciocchi, R., 191
Bacterial activity, 11
    bioleaching, 195
    in bioremediation processes, 55, 57
    sediment quality and, 90
    sulfate-reducing, 60
Banta, G., 61

Barcelona Convention, 171
Barium chloride, 48
Barriers to technology development and
    implementation, 253
Bedard, D. L., 155
Belgium, 284
Beneficial uses
    of dredged materials, 243–246
    of sediments, 196–198
Benthos, 30–31, 90
    natural recovery and, 109, 113
    steel slag and, 160
Bentley, S., 46
Benzenes and BTEX, 53, 88–89, 126
Berg, U., 137
Bioaccumulation of contaminants, 12–14, 59–60
Bioassays, 91
Bioattenuation, 58–59
Bioaugmentation, 154, 155, 156
Bioavailability, 58–59, 91
Bioconversion processes, 195
Biogenesis, 176
Bioleaching, 195
Biological remediation technologies, 56–58,
    153–156, 190–196
    case studies, 161–163
Bioreactors, slurry, 191
Biostimulation, 154
Biosurfactants, 176–178, 191, 204
Biosurveys, 91
Biotransformation and degradation of organic
    chemicals and heavy metals, 54–59
Bioturbation, 60–62
    natural recovery and, 109, 112, 116
Birge-Ekman samplers, 76
Bloom, N. S., 49
Brackish sediments, 21–22, *103–105*
Bray, R. N., 170, 247, 249
Brenner, R. C., 126
Bridges, T. S., 215
Brummer, G. W., 48
Bucalá, V., 189
Burning temperatures, 29

C

Canada, 215, 279
    disposal at sea in, 242
    dredging in, 173–174
    Lachine Canal, Montreal, 249–253
    sustainability evaluation in, 247–248